D1794700

NUCLEAR ENERGY
PROMISE *or* PERIL?

NUCLEAR ENERGY
PROMISE *or* PERIL?

Executive Editor

B. C. C. van der Zwaan

Institut Français des Relations Internationales, France

Advisory Editors

C. R. Hill

Institute of Cancer Research, University of London

A. L. Mechelynck

Huldenberg, Belgium

G. Ripka

(formerly) Centre d'Etudes de Saclay, France

Published by

World Scientific Publishing Co. Pte. Ltd.

P O Box 128, Farrer Road, Singapore 912805

USA office: Suite 1B, 1060 Main Street, River Edge, NJ 07661

UK office: 57 Shelton Street, Covent Garden, London WC2H 9HE

British Library Cataloguing-in-Publication Data
A catalogue record for this book is available from the British Library.

NUCLEAR ENERGY
PROMISE OR PERIL?

ISBN 981-02-4011-2

Printed in Singapore by Uto-Print

A Peer Review Workshop of the
Pugwash Conferences on Science and World Affairs
on "The Prospects of Nuclear Energy"
held in Paris, December 4–5, 1998

Table of Contents

Preface

Energy is one of mankind's basic needs. The production of energy has side-effects that can be widespread and even global in their impact. Whilst being a necessary ingredient for development, energy constitutes a potential source of hazard, tension and conflict. We are far from having achieved a responsible approach to energy production, which would take into account the impact on the environment and the exhaustion of reserves of cheap non-renewable fuels.

Today, nuclear reactors provide about 17 % of electrical power, world-wide, but with a very unequal distribution geographically. Governmental policies regarding nuclear power are equally varied and are influenced by public opinion as well as by the availability and price of alternative energy sources. As a result, the long term future of nuclear energy (up to about the year 2050) is far from clear. It has some obvious advantages: nuclear reactors do not produce CO_2 and thus nuclear energy does not contribute appreciably to the "greenhouse effect", nuclear fuel is far more evenly distributed geographically than are oil and coal, and new technologies are being developed (breeders and hybrid reactors) which make far better use of fissile materials, and which can reduce the effective lifetimes of nuclear wastes. However, the safety of at least some current designs of reactors is still strongly open to question. In addition, nuclear wastes need to be stored for long periods of time, entailing, in the long term, both cost commitments and risks of human and environmental damage. Perhaps most importantly, there always remains a possibility that the fissile materials generated by reactors could be diverted to facilitate the proliferation of nuclear weapons.

There are thus strong arguments for moving away from reliance on nuclear energy. However, the consequences of doing this must also be assessed quantitatively, particularly in the light of the considerable increase of energy needs in developing countries.

The nuclear industry is characterised by long intervals of time (of about 20 years) between the conception and the industrial realisation of new technologies. Since the current generation of nuclear reactors will provide electricity up to the years 2010-2020, it is now timely to examine the environmental and security issues which can bear on the long term future of nuclear energy.

It was against this background that plans were made for writing a book that would provide authoritative accounts of some of the main issues of topical interest, and would do so in a manner that is accessible to the non-expert. As a key part of the publication process all the authors attended, together with an additional group of scientists and others from a wide range of disciplines, a two-day "workshop" at which all draft texts were presented, discussed and criticised - thus helping the authors to ensure that their contributions were fair and accurate. The workshop meeting was organised jointly by the French, Belgian and British Pugwash[1] groups, but ultimate responsibility for factual

[1] Pugwash Conferences on Science and World Affairs.

accuracy, and for the opinions expressed, is entirely that of the individual authors and of the editors.

The contents of the book fall broadly into three parts, dealing (respectively, and in shorthand terminology) with Energy, Health and Weapons Proliferation.

The prospect of a world energy crisis is widely appreciated and, in early chapters, we review the background to this in some detail, particularly exploring the role in it, if any, that nuclear energy may be called upon to play. An important factor here is the very different needs and perceptions in developing countries, as contrasted with highly industrialised countries.

Much of the concern expressed about nuclear energy, particularly in North America and in some European countries, has focused on the risks posed to human health by both routine and accidental releases of radioactivity to the environment. In a series of chapters we therefore take a look at the problems involved in ensuring acceptably safe operation of power generating reactors, and in management of their eventual waste products.

The other major cause for concern, which many feel to be particularly worrying, is the production by reactors of nuclear explosive material: plutonium. The concluding section of the book therefore assesses the nature of this problem and the options for addressing it. At least for the immediate future this must entail international collaboration on devising and implementing effective administrative safeguards. In the longer term, however, it will clearly be very desirable to reduce, if not eliminate, global stocks of these materials, and we therefore also discuss some ideas on how this might be done.

Some of the topics that are dealt with in the book are technically somewhat complex. We have therefore provided an introductory chapter that may help the non-expert to follow the subsequent, more detailed discussions.

This book makes no pretence to provide comprehensive treatment of what is, of course, an enormous subject. Nor, even, can it deal fully with all the issues that are of potential public concern. If, however, it stimulates and assists informed discussion of some of the more pressing issues that will need to be resolved in the coming years, we shall feel that it has served its purpose.

Kit Hill
André Mechelynck
Georges Ripka
Bob van der Zwaan

Chapter 1 Nuclear Electricity - An Aide Memoire

by
C.R. Hill and R.S. Pease*

Introduction

The purpose of our project is to assess the role of nuclear energy technology in the context of achieving a global economy that is, at the same time, politically secure and environmentally safe and sustainable. Our report is aimed at a broad group of readers, not all of whom may have background knowledge of nuclear science and technology sufficient to follow some of the detailed arguments that are discussed. It is for such readers that the present chapter is intended: experts may prefer to skip. In attempting to present a large and somewhat complex subject in concise form, it has been necessary to be selective and, to some extent, simplistic. Fuller accounts are, of course, given in other published texts. Two that seem particularly useful, and on which we have drawn considerably in preparing this chapter are those by Price (1990) and Wilson (1996), and these are now updated by summaries of a recent international symposium (IAEA, 1998).

1. Basics

The source of the energy produced by present-day nuclear power stations is attributable to three related phenomena:

(1) The potential of certain heavy atomic nuclei for *fission*: splitting into two parts of roughly equal mass *(fission products)*, together with a few free neutrons. Such nuclei are referred to as *fissionable*. (Nuclei that are fissionable by interaction with *thermal* neutrons - see below - are referred to as *fissile*).

(2) The potential of neutrons thus released to initiate a fission event in one or more neighbouring fissionable nuclei, resulting in a *chain reaction*.

(3) The very substantial release of free energy that occurs in the fission process. This release of energy is reflected by an equivalent deficit in mass as between a fissioned atom and the sum of the products of its fission: the two daughter atoms together with the surplus neutrons.

---

* Kit Hill is Secretary of the British Pugwash group and formerly Professor of Physics as Applied to Medicine at the Institute of Cancer Research, London.
Bas Pease is Chairman of the British Pugwash Group and formerly Director of Fusion Research, UK Atomic Energy Authority.

The mass-deficit phenomenon is illustrated in figure 1, from which it will be seen that, for example, fission of a nucleus of atomic number 92 (uranium) into two fragments, which will typically be in the range 40-60, would entail a reduction in total isotopic mass.

Figure 1 also illustrates the existence of an additional potentially energy-releasing phenomenon, which we shall for the moment simply note in passing: that entailing the combination, or *fusion* of two or more very light nuclei, e.g. ^{2}H (deuterium), to form a single heavier nucleus, e.g. helium.

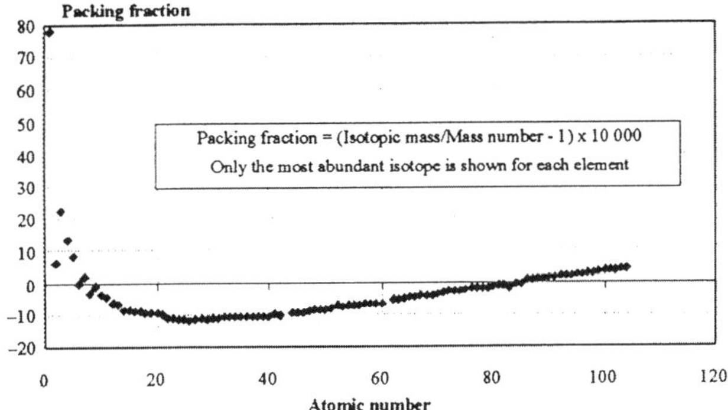

Fig. 1. Variation with atomic number of mass deficit (packing fraction), normalised to zero, for ^{14}C, by definition of atomic mass units (amu). *Source: from Wilson, 1996.*

The energy released in the process of fission is very large: some 200 MeV per fission, thus yielding roughly a million times more energy than would come from combustion of the same mass of coal. Put another way, the energy released by fission of 1 kg of U-235 amounts to some 60 TJ When converted to electricity with 30 % efficiency, typical of most nuclear power reactors, this heat will generate some 5 GWh of electricity.

In exploiting these phenomena for the production of heat, and thus electricity, the central engineering problem is one of control: the release of energy needs to take place at a rate that can be controlled according to requirements of power generation; the process must never build up dangerously in a manner that escapes such control, even under circumstances of machine fault or human error; and the system must continue to run in this way for periods of many years. The factors that enable such control to be achieved include the following.

(1) The probability that a neutron will induce fission on collision with a fissionable atom (technically the *cross section* for that process) depends strongly on its kinetic energy, or in other words its speed. The energy of the neutrons released on fission is about 1 MeV (i.e. they are *fast*), whereas the fission cross section has relatively much higher values for *slow* neutrons (specifically those that are in approximately *thermal* equilibrium with their surroundings).

(2) To exploit this behaviour the (fast) fission neutrons may need to be slowed down, i.e. *moderated.* This can be achieved by arranging for them to undergo multiple elastic collisions with atomic nuclei of comparable mass; ideally hydrogen, but carbon and deuterium are also used.

(3) Typically, each fission event releases some 2.5 neutrons. For a sustained chain reaction in a steady state (as ideally required for power generation) exactly one of these must induce a further fission. This is the so-called state of *criticality*. In general, the average number of neutrons from each fission that is effective in propagating the chain (k_{eff}) may be greater or less than unity.

(4) Evidently, if k_{eff} is persistently greater than unity the chain reaction will accelerate exponentially, with eventually explosive outcome. Correspondingly, if it is persistently less than unity, the process will slow down and stop. Thus, as in all control situations, there is a requirement for *feedback* mechanisms that will sense excursions from $k_{eff} = 1$ and introduce appropriate corrections that will maintain the required steady state. In practice, as will be seen, a number of feedback processes, both positive and negative, may be operative in any given situation, some specifically engineered and others inherent in the underlying physics.

(5) A factor that contributes vitally to negative feedback is removal of neutrons from the system by absorption following *inelastic* collisions with a wide variety of nuclei. This is the function of *control rods* in a reactor. In another guise, this factor limits the operational life of reactor fuel because of the build-up of neutron-absorbing fission products (in this context referred to as *poisons*).

(6) More than 99 % of fission neutrons are emitted at the instant of fission: so called *prompt neutrons*. The remainder, *delayed neutrons* (0.7 % for U-235 and 0.2 % for Pu-239), are emitted by radioactive decay of several different fission products having half-lives varying from 0.05 to 56 s. This component of delayed neutrons is vitally important as a safety mechanism, in allowing adequate time for servo-operated control rods to respond to small excursions from equilibrium.

(7) Absorption of neutrons following collisions with neighbouring nuclei results in *transmutation* of the nuclei to other types, with corresponding formation of *activation products*, which themselves are generally radioactive.

(8) The heat released in radioactive decay of the fission and activation products that will have built up during normal operation constitutes about 6 % of the total fission power. This is of substantial practical importance and must be allowed for both in the event of sudden shut-down of a reactor and in avoiding overheating of spent fuel following its removal from a reactor.

2. Thermal Reactors

There are five types of nuclear reactor currently in commercial use world-wide, all of which exploit interactions of thermal neutrons. Of these five, three employ water as a moderator, principally in the form of the so-called *pressurised water reactor* (PWR). We shall therefore use this as our model for a main description of the functioning of a reactor, and then note how other types differ from this.

2.1 The Pressurised Water Reactor (PWR)

The principles of operation of a PWR are illustrated in figure 2. The uranium oxide fuel is packaged, as *fuel rods*, into sealed hollow cylinders formed from a zirconium alloy, zircaloy, so chosen because of its low neutron absorption cross-section relative to stainless steel. For this type of reactor, the uranium itself needs to be enriched in its U-235 content to some 4 % (compared with its natural abundance of around 0.7 %), in order to achieve a workable value for k_{eff}. An array of these fuel rods is placed inside a pressure vessel through which there is a constant flow of water that transfers heat to a set of, usually three or four, heat exchangers that in turn generate steam to drive turbines. Interlaced with the array of fuel rods is a set of *control rods*, containing a neutron absorber (such as hafnium, boron or cadmium), whose position can be continually adjusted to maintain a dynamically stable operating condition. In emergency, or for refuelling, additional control rods can be rapidly inserted in order to achieve shut-down. The operating temperature is typically around 300 C and pressure is maintained high enough to prevent boiling. The whole assembly is contained within a heavy concrete shield, designed to be capable of withstanding the consequence of any possible rupture of the pressure vessel under normal operating conditions. Refuelling entails shutting down the reactor for a period of a few weeks, opening the pressure vessel, and replacing some 30 % or more of the fuel rods on each occasion.

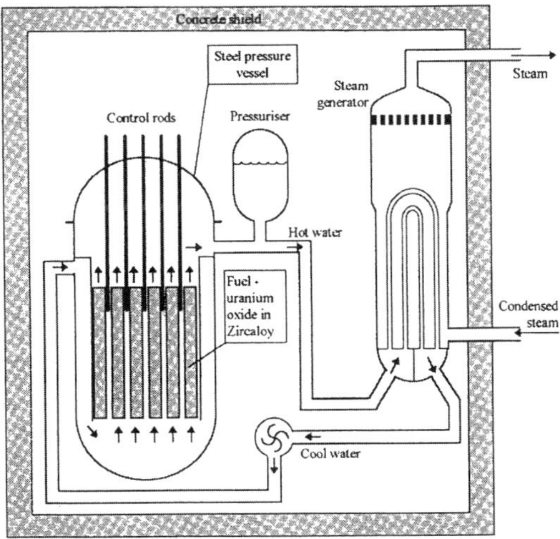

Fig. 2. Simplified layout of a pressurised water reactor.[2] *Source: Wilson, 1996.*

[2] Note, e.g., that in practice a number of heat exchangers are used in parallel.

2.2 The Boiling Water Reactor (BWR)

The BWR is, in principle, very similar to the PWR, with the difference that the water is under lower pressure (some 70 instead of 150 bar) and is allowed to boil, thus providing a direct source of steam for the turbines. As a consequence, however, the turbines are part of the reactor circuit and, as the coolant in the circuit is inevitably contaminated with radioactivity, this leads to offsetting problems in shielding and maintenance costs.

2.3 Heavy Water Reactors

The only design of this type in routine use is the CANDU (Canadian Deuterium Uranium) system. This uses heavy water, D_2O, as the moderator which, although expensive, allows the use of natural uranium oxide fuel. It is a pressure-tube reactor, with the fuel elements located in horizontally arranged tubes through which heavy water coolant is circulated under pressure. In order to minimise risk of leakage of tritium activation product into the steam turbines, the cooling circuit is separated from the moderator D_2O, which is not pressurised. Refuelling is carried out "on load" (i.e. whilst generating electricity), entailing a particularly high requirement for safeguards against illicit diversion of fissile material. The use of separate cooling and moderator circuits results in the lowest steam temperature, and thus lowest thermodynamic efficiency, of all the major reactor designs. Against this, there is claimed to be a benefit in that the substantial bulk of coolant constitutes a large heat sink which, in combination with other features of the design, should contribute to safety.

2.4 Gas Cooled, Graphite Moderated Reactors

Also driven originally by a need to avoid reliance on enriched uranium fuel has been the development of a number of types of reactor that use graphite as an efficient, if bulky moderator, and carbon dioxide gas as a coolant. The so called Magnox reactors (using a magnesium alloy, resistant to corrosion and of very low neutron absorption for encasing the fuel) were developed in the UK on this principle, and were then followed there by the Advanced Gas-cooled Reactor (AGR) design. This employs low-enriched fuel in stainless steel cans and runs at a higher temperature than the PWR, with correspondingly higher thermodynamic efficiency. Plans, and test construction, are now afoot in China for a further development of the gas reactor principle with target operating temperatures of 950 C.

2.5 Water Cooled, Graphite Moderated Reactors (RBMK)

The RBMK (*Reaktor Bolshoi Moshnost Kanaly*, i.e. "high power channel reactor") design originated in the former Soviet Union (FSU) and found widespread use there, including at Chernobyl. Its fuel elements use low-enriched uranium and are located in separate channels containing water that is under pressure of 78 bar but allowed to boil at some 290 C, with the steam directly driving turbines, as in the BWR design. The safety of the RBMK reactor has been exhaustively analysed since the Chernobyl accident. In this, the reactor went supercritical and the control mechanisms then failed to prevent development of a power surge, which was halted only by the explosive disassembly of the whole core (Novikov *et al.,* 1995).

3. Radioactive and Fissile Products

As already mentioned, radioisotopes originate in a reactor in two ways: as fission products and as activation products (generated by the absorption of neutrons). In a well behaved reactor, with no unscheduled leaks, fission products should be confined to where they arise: within the fuel rods. Activation products however arise wherever neutrons may penetrate, which means in practice throughout the reactor and its shield.

Fission of isotopes such as U-235 or Pu-239 typically results in a bi-modal (double-peaked) distribution of abundance of fission products, in the manner illustrated in table 1. The pattern of initial abundance of activation products is, by contrast, very variable, and is dependent primarily on the atomic/isotopic composition of the neutron-irradiated structures within the reactor and of corrosion products in the water.

Eventual abundances of radioisotopes within a fuel rod, or at any other location in a reactor, will in general be very different from these initial abundances. They will be strongly influenced by decay of the radioisotopes, at varying half-lives, and also by multiple transmutation events experienced by single nuclei. Thus, in the management of these end-products, different groups of radioisotopes will be of predominant importance in different contexts and at different points in time. In the event of shut-down of a reactor, for example, because they are releasing energy very fast, radioisotopes with half-lives of the order of seconds and minutes will be of concern as a potential cause of overheating. At the other extreme, those with half-lives of hundreds of years and more present potential problems in long-term disposal.

A further set of radioisotope characteristics that is of importance in their management is the nature of their radioactive emissions: α, β and γ-rays, and the energies thereof. Those emitting high energy γ-rays, for example, present a problem in terms of the need for protective shielding whilst they are manipulated and transported, whilst α-particle emitters tend to be particularly toxic biologically. The decay and radiation characteristics of a number of the more important radioisotopes in this context are summarised in table 1. More detailed data are given by Schapira - this volume (see table 1).

During its residence in a reactor, a fraction of the U-235 content of the fuel elements is converted to fission products. Most of these are radioactive, and most behave as solids, although some, such as Kr-85, are noble gases. These fission products proceed to decay, generally into other radioactive species, and/or to be transformed to a different isotopic form following capture of neutrons (*activation*). Another, particularly important group of activation products is that resulting from neutron capture in the U-238 that constitutes the bulk of the content of the fuel elements. Some of these activation products (or, in turn, their radioactive decay products), and notably Pu-239, are themselves fissile. In this sense - that neutron activation induces them to *breed* fissile isotopes - U-238 and other isotopes behaving in this way are said to be *fertile*.

With continued exposure to neutrons in a reactor, the Pu-239 is itself transformed to higher-mass isotopes, and particularly Pu-240, which is not fissionable by thermal neutrons. In fact, for the purposes of constructing an explosive nuclear weapon, Pu-240 turns out to be a harmful contaminant. For this reason, reactors being operated for the production of *weapons grade plutonium* employ sufficiently short fuel cycle times that only small quantities of Pu-240 and similar contaminants have time to build up. Relatively pure Pu-239 can thus be recovered. It is important to note here, however, that even *commercial grade plutonium*, i.e. with substantial Pu-240 and higher contamination, could be used to construct a weapon, although possibly of reduced power and reliability.

The major component of the *spent fuel,* that remains after its residence in a reactor is complete, is uranium. As compared with its initial composition, this will necessarily be somewhat depleted in U-235 but, in reactors using enriched uranium, the content of the spent fuel may remain substantially enriched, with corresponding potential economic value. Other major constituents of spent fuel will be fission products and activation products, the latter including in particular isotopes of plutonium and higher *actinides* (members of the chemical group that includes uranium and plutonium). Some of these are also listed in table 1.

There are broadly two options for dealing with spent fuel: store it indefinitely as is; or carry out a chemical separation process ("*reprocessing*") aimed at extracting certain wanted and/or very long-lived components (e.g. uranium, plutonium) and then store the residue. The former option is considered to be the more secure (although only in the short run), on the grounds that it maintains, for a substantial period, a high level of penetrating radiation emission, thus inhibiting undesirable availability of plutonium for weapons use. Indeed, a favoured option for secure disposal of surplus weapons plutonium would be to mix it with spent fuel, thus attaining the so called *spent fuel standard* of security.

Element	Mass Number	Half-life	Radiation Emission [(a)] (type; energy, MeV)	Fission Yield (%)	"Dose Coefficients" [(b)] Ing	Inh
Products of fission of U-235						
Kr	85	10.7 y	β, γ ; 0.69	0.3	n.a.	n.a.
Rb	87	$4.8\ 10^{10}$ y	β ; 0.26	2.5	$1\ 10^{-3}$	$5\ 10^{-4}$
Sr	89	50.5 d	β ; 1.46	4.8	$3\ 10^{-3}$	$6\ 10^{-3}$
Sr	90	29.1 y	β ; 0.55+2.3	5.8	$3\ 10^{-2}$	$4\ 10^{-2}$
Y	91	58.5 d	β ; 1.55	5.4	$2\ 10^{-3}$	$7\ 10^{-3}$
Zr	93	$1.5\ 10^{6}$ y	β ; 0.06	6.5	$1\ 10^{-3}$	$1\ 10^{-2}$
Zr	95	64 d	β, γ ; 1.1+0.92	6.2	$1\ 10^{-3}$	$5\ 10^{-3}$
Mo	99	2.75 d	β, γ ; 1.38	6.1	$6\ 10^{-4}$	$1\ 10^{-3}$
Ru	103	39.3 d	β, γ ; 0.76	3.0	$7\ 10^{-4}$	$2\ 10^{-3}$
I	129	$2\ 10^{7}$ y	β ; 0.15	0.8	$1\ 10^{-1}$	$2\ 10^{-2}$
I	131	8.04 d	β, γ ; 0.97	3.1	$2\ 10^{-2}$	$2\ 10^{-3}$
Te	132	3.3 d	β, γ ; 0.41	4.7	$4\ 10^{-3}$	$2\ 10^{-3}$
I	133	20.8 h	β, γ ; 1.8	6.9	$4\ 10^{-3}$	$6\ 10^{-4}$
Xe	133	5.2 d	β, γ ; 0.43	6.6	n.a.	n.a.
Cs	135	$2.3\ 10^{6}$ y	β ; 0.21	6.4	$2\ 10^{-3}$	$3\ 10^{-3}$
Cs	137	30.2 y	β, γ ; 1.18	6.2	$1\ 10^{-2}$	$1\ 10^{-2}$
Ba	140	12.7 d	β, γ ; 1.05+3.48	6.4	$3\ 10^{-3}$	$5\ 10^{-3}$
Ce	141	32.5 d	β, γ ; 0.58	6.0	$7\ 10^{-4}$	$3\ 10^{-3}$
Ce	143	1.4 d	β, γ ; 1.44+0.93	5.7	$1\ 10^{-3}$	$8\ 10^{-4}$
Ce	144	284 d	β, γ ; 0.32+3.01	6.0	$5\ 10^{-3}$	$4\ 10^{-2}$
Nd	147	11 d	β, γ ; 0.91+0.23	2.7	$1\ 10^{-3}$	$2\ 10^{-3}$
Activation Products						
H	3	12.6 y	β ; 0.02	-	$2\ 10^{-5}$	$5\ 10^{-5}$
C	14	$5.5\ 10^{3}$ y	β ; 0.16	-	$6\ 10^{-4}$	$2\ 10^{-3}$
Co	60	5.3 y	β, γ ; 2.8	-	$3\ 10^{-3}$	$1\ 10^{-2}$
Pu	239	$2.4\ 10^{4}$ y	α ; 5.15	-	$3\ 10^{-1}$	$5\ 10^{+1}$
Pu	240	$6.6\ 10^{3}$ y	α ; 5.17	-	$3\ 10^{-1}$	$5\ 10^{+1}$
Pu	241	14 y	β ; 0.02	-	$5\ 10^{-3}$	$9\ 10^{-1}$
Pu	242	$3.8\ 10^{5}$ y	α ; 4.9	-	$2\ 10^{-1}$	$5\ 10^{+1}$
Am	241	433 y	α ; 5.49	-	$2\ 10^{-1}$	$4\ 10^{+1}$
Np	237	$2.1\ 10^{6}$ y	α ; 4.79	-	$1\ 10^{-1}$	$2\ 10^{+1}$
Fuel						
U	235	$7.0\ 10^{8}$ y	α ; 4.40	-	$5\ 10^{-2}$	3
U	238	$4.5\ 10^{9}$ y	α ; 4.20	-	$5\ 10^{-2}$	3
Th	232	$1.4\ 10^{10}$ y	α ; 4.0	-	$2\ 10^{-1}$	$5\ 10^{+1}$

N.B. This table is illustrative only - not comprehensive. Radioisotopes included in the table are those whose combination of expected abundance and half-life make them important candidates for possible irradiation of the general population following accidental or other dispersal of reactor material.

(a) Including emissions of daughter products, where appropriate.

(b) "Dose coefficients" given here, for ingestion (Ing) and inhalation (Inh) respectively, and expressed in microsievert per becquerel, are those calculated for an adult, to age 70 years (ICRP, 1996). Whilst inevitably being somewhat arbitrary parameters, such coefficients serve to illustrate the relative toxicities of the radioisotopes as internal emitters. A further contribution to dose may arise from external irradiation, mainly from γ-emitters.

Tab. 1. Characteristics of some of the more abundant radiotoxic isotopes in thermal reactors.

4. Health Hazards

The radiations emitted by radioisotopes associated with nuclear reactors have the potential, in the course of their life times, to cause damage to humans and other living organisms. Damage to humans can arise broadly in two ways: from emissions originating outside and inside the body respectively. The effectiveness of the former arises predominantly from emissions of penetrating γ-rays, whilst that of internal emitters depends critically on the chemical nature and form of the radioisotope, and thus the manner and extent to which it participates in metabolism and deposits within range of a radio-sensitive site. It is useful to distinguish two classes of radiation effects in humans:

(1) *Acute, deterministic effects.* Here clinically observable effects (e.g. "radiation sickness", death) generally appear within a few weeks, although certain effects such as cataracts may have a latent period of some years. These effects are subject to a threshold in dose, in the sense that there is a dose below which no effect will be found. (For example, a human whole-body dose above 5 sievert - see below - is almost invariably lethal, as a consequence of damage to the intestine, whilst below 1 sievert clinical symptoms are generally limited to occurrence of radiation sickness in just a small proportion of the exposed population.) In general, effects in this category are specific to radiation injury.

(2) *Chronic, stochastic effects.* This group comprises genetic changes and cancer induction, which is generally observed only after a latent period of some 20 years or more (except for some childhood cancers and leukaemia, which may occur some 3-5 years after exposure). Genetic effects may only express themselves after several generations. Radiation-induced cancers are of the same types that occur in the general population spontaneously, generally due a combination of several causes (smoking, diet, exposure to some chemicals or environmental agents, genetic factors, and others as yet unidentified). For this reason, attribution of observed cancers to radiation has generally to be done on a statistical basis. The relationship between cancer incidence or death and radiation dose is derived from epidemiological studies, in which the exposed population is compared with a suitably matched control population. The most accurate data are those that have been obtained from studies of survivors of the Hiroshima and Nagasaki bombs. At high doses (above 100 millisievert) there appears to be a linear relationship between excess cancer deaths and whole-body radiation dose. From the point of view of possible hazard to the general, and most of the working population, from exposures due to nuclear power and medical procedures, the main interest is in the effect of small doses (less than 10 millisievert). In this range, however, the error of the observations is usually too large to yield statistically significant values, and resort to theoretical models becomes necessary. Apart from the linear model, two other models have been put forward. One, which is supported by some limited experimental evidence, is the sub-linear model (e.g. a quadratic increase with dose), in which the effect at very low doses is so small that, in practice, one can speak of a threshold, or safe dose, below which there is no significant effect. In contrast, a supra-linear model has been suggested, in which the effect of small doses would be proportionally greater than that predicted by the linear model.

The magnitude of biological damage that results from a radiation exposure of a "target tissue" may depend on a number of factors, including: the radiation energy absorbed; the nature of the radiation delivered (i.e. α, β, γ); the rate at which it is delivered; and the inherent *radiosensitivity* of the target tissue.

The *absorbed dose* due to a radiation exposure (energy absorbed from the radiation beam per unit mass of absorber) is expressed in *gray* (Gy ≡ joules/kilogram). In calculating a *biologically effective dose*, in relation to a particular effect, it is necessary to multiply the absorbed dose by a *quality factor* appropriate to the particular radiation involved. The result is expressed in *sievert* (Sv). Quality factors are taken relative to the effectiveness of X-rays and are approximately unity for γ-rays and β-particles, but generally in the range 5-20 for α-particles and fast neutrons. In illustration of the magnitudes of radiation doses commonly encountered, that delivered *to a target tumour* in a course of radiotherapy will be some 50 Sv, whilst the annual whole-body natural background dose experienced by typical humans is some 2.4 millisievert (mSv); in addition to which there is a contribution that is generally in the range 0.3-1.0 mSv per annum from all man-made sources - predominantly from diagnostic X-rays (UNSCEAR, 1988; see also the appendix of this book). The current international recommendations for upper limits on annual radiation doses received by occupational workers and the general population are 20 and 1 mSv respectively (ICRP, 1990).

To put this in perspective, the United Nations Scientific Committee on the Effects of Atomic Radiation (UNSCEAR, 1988) estimates the lifetime risk for an adult of suffering a radiation-induced lethal cancer to be in the range 4-11 10^{-5} milligray of whole body dose, with a somewhat higher risk for a child. In addition, the incidence of detectable genetic abnormalities in children of exposed individuals is estimated (based on animal data, in the absence of human data) to be of the order of 10^{-5} per milligray dose to gonads, with additional incidence occurring in subsequent generations.

The occupational hazards associated with work in the nuclear industry is a complex subject that cannot be covered here. In general however, in the absence of drastic accidents, doses are maintained within internationally recommended limits. Historically the most dangerous procedure, and the only one for which there seems to be positive epidemiological evidence of pathology, has been uranium mining.

The most credible source of irradiation of the general population would be fuel elements, either because of an uncontained leakage failure during reactor operation, or later, following either accidental or deliberate dispersal of material to the environment in the course of storage, disposal and/or reprocessing. In the former situation, gaseous and volatile species (e.g. Kr-85 and I-131) tend to be preferentially released, as was the case in the 1957 UK Windscale accident. In the latter situation, isotopes of particular concern are likely to be those that have "longish" half-lives (see table 1) and are also chemically mobile. It will be "highish" energy γ-emitters that may contribute to external dose, whilst isotopes of elements that have common metabolic analogues (e.g. Sr-90, Ca, Cs-177, K) are potential internal emitters. It is noteworthy in this connection that most such radioisotopes have commonly occurring natural radioactive counterparts that exhibit similar metabolic behaviour (Sr-90, Ra-226, Cs-137, K40, and even Pu-239, Th-232), the principal exception being the iodine isotopes. The annual *per caput* dose, averaged over the world population, as a result of nuclear power activities, is currently estimated to be 0.15 microsievert (UNSCEAR, 1988).

There are substantial bodies of both laboratory animal and human epidemiological data on which current assessments of radiation health hazards are based. The latter category includes studies of mining, industrial and professional occupational exposures together with, outstandingly, evidence from the very large populations exposed to sub-lethal radiation doses at Hiroshima and Nagasaki, and more recently Chernobyl. Whilst there are inevitable deficiencies in such data, particularly in dosimetry and in baseline disease incidence, when set beside animal evidence they provide good grounds for drawing conclusions as to the likely nature and quantitative limits to expected hazard from given exposures. Such conclusions are summarised and substantiated in a number of authoritative publications, such as those of the International Commission on Radiological Protection (ICRP) and UNSCEAR.

The potential health risks to the public that could follow from a fission power reactor accident arise from four main circumstances: (i) emissions from the very large amount of radioactive material contained in the reactor at any given time; (ii) the energy potential of the fuel in the reactor - of the order of one year's heat output; (iii) the potential for the neutron chain reaction operating within the entire mass of fuel to diverge sufficiently that containment is ruptured before the reaction is halted by disassembly of the core; and (iv) persistence of heat production in the fuel elements after the reactor is shut down so that, if the cooling system fails, meltdown will occur. So far, two accidents affecting the general public have occurred to civil nuclear power stations: that at Chernobyl in 1986 was due to cause (iii) above, whilst that at Three Mile Island in 1979 was due to cause (iv). Modern nuclear power stations are designed to meet safety criteria laid down by national regulatory bodies. In the US, the Nuclear Regulatory Commission (NRC) requires, *inter alia,* that the mean frequency of a large release of radioactive materials from a reactor accident should be less than one in one million per year of reactor operation. Similar criteria apply in Europe. The UK defines a serious accident as one that might subsequently give rise to 100 deaths from cancer (Roberts *et al*., 1990).

The health consequences of the Chernobyl accident have been intensively studied and a detailed review of the evidence to date, taking account particularly of information accumulated by the World Health Organisation (WHO), UNSCEAR and other agencies, was compiled on the tenth anniversary of the event (OECD, 1995), whilst a very useful analysis of the reasons for the accident has been given by Novikov *et al* (1995). In outline, the Organisation for Economic Co-operation and Development (OECD) conclusions are the following:

(1) 31 reactor and emergency workers died within about a month of the accident as a result of either blast, burns, or acute radiation injury. Some 200 more such workers exhibited clinical signs of acute radiation syndrome, from which they subsequently recovered.

(2) No evidence of such acute radiation effects was found amongst non-occupationally-exposed members of the public.

(3) The one physical health consequence for which there was strong direct evidence amongst the general public was an abnormally high, dose-related incidence of thyroid cancer, particularly in children. In the regions most seriously affected (Belarus, Northern Ukraine and parts of Russia), for the period 1986-94, 847 cases of childhood thyroid cancer (with three deaths from the disease) were recorded, as against an expected incidence of 54 on the basis of pre-1986 records. Further increased incidence is to be expected, at least for the next ten years and, although the disease is generally treatable with substantial success, the predominant form in this epidemic (an aggressive papillary carcinoma) has proved abnormally difficult to control.

(4) A number of epidemiological studies have been carried out, in populations both from the FSU and elsewhere, of possible links between Chernobyl-related radiation exposure and other forms of cancer, particularly childhood leukaemia, but all of these hitherto have shown no elevated incidence of disease.

(5) At the same time, however, radiation doses measured not only for those living in the vicinity of Chernobyl, but also for substantial populations living elsewhere in the Northern Hemisphere, would imply (on the assumption of a linear dose extrapolation), an 0.5 % lifetime excess incidence of all cancers for people living in the "control zones" around Chernobyl, and a corresponding 0.01 % excess averaged over the European population as a whole.

(6) Whilst there were numerous anecdotal claims of other physical ill-effects of the accident (some leading to apparently formal diagnoses, such as "Vegetative Dystonia"), none of these has been substantiated in systematic studies. At the same time however, the immense psychological and sociological trauma that resulted was very real indeed. This seems to have been influenced by a number of factors: the magnitude of the accident itself; the manner in which the consequences for local people, and particularly the evacuation, were managed by the authorities; and a general perception of the frightening, because mysterious, nature of exposure to radiation.

5. Further Developments

There is a number of developments currently in progress, or being studied, that may have implications for the future viability and acceptability of nuclear power. Briefly, these are as follows:

5.1 "Enhanced Safety" Reactor Designs

Much of the concern about safety of current reactor designs centres on three issues: reliance on externally powered pumps and valves for emergency close-down of the reactor; the possibility of breaks or leaks in high-pressure pipework located outside the main pressure vessel; and the possibility of mechanical failure at the points where pipework penetrates the pressure vessel. Such considerations have resulted, for example, in proposals for a so-called "Safe Integral Reactor" variant of the PWR, in which all primary circuit components, including steam generators, are situated within the pressure vessel, whose large size also helps to provide passive cooling of the core in the event of a malfunction.

5.2 Plutonium Fuelled Thermal Reactors: MOX

The need for fissile-enrichment of the uranium fuel burnt in most current reactors can be met, in principle, by addition of fissile isotopes of plutonium, in practice in the form of *Mixed Oxides* of plutonium and uranium: "MOX". This recycling approach increases, approximately from 2 % to 3 %, the proportion of uranium energy that can be utilised. Whilst this technique is now being exploited on a substantial scale (and has obvious attractions as a means towards elimination of weapons-usable plutonium), there is an important limitation. As compared with uranium, plutonium has substantially higher cross sections for both fission and absorption, with the consequences, first, that there will be fewer neutrons available to sustain the chain reaction (technically a lower neutron multiplication factor) and, second, that there will be increased competition for neutrons with the neutron-absorbing reactor control mechanisms. Thus, reactor design will need to take into account the use of MOX fuel and, in with current technology, the proportion of such fuel loaded in a reactor is limited to some 30 %.

5.3 Fast Reactors

Although, as was described above, the cross sections for inducing fission in U-235 or Pu-239 are relatively low for fast neutrons, it is nonetheless possible to design a reactor to run without a moderator, at some cost but also with some important features. An important part of the "cost" side is the need to use a highly effective but non-moderating coolant fluid. For this, both molten lead and molten sodium have been used experimentally; in each case however substantial, and so far incompletely solved engineering problems have been encountered. Much of the original interest in fast reactors arose from their potential (although not necessity) of functioning as *breeders* of more fissile material than they consume (see Eyre *et al.*, 1990). They are thus able to exploit the energy potential of (fertile, but not fissile) U-238, whose natural abundance is some 140 times that of U-235. Practically, therefore, the neutron-generating, active core of a reactor is surrounded by a heavy blanket of natural, or even depleted uranium, in which future supplies of fuel are generated. An essential requirement, however, is that the primary fuel releases, on fission, a sufficient number of neutrons to both sustain the reaction and provide for breeding and, for this reason, the main interest hitherto has centred on the use of plutonium as a fuel (all of whose isotopes are fissionable by fast neutrons) - one of the primary justifications adduced for the commercial reprocessing of thermal reactor fuel. Fast reactors, however, have proved to be technically difficult and relatively expensive, and their development has ceased in many countries.

5.4 Thorium Reactors

It is also possible, in principle, to exploit the energy content of thorium-232, whose natural abundance is some 3-4 times that of uranium and which can occur locally in rather high concentrations, notably in India, which has little uranium. Fuel consisting of mixed oxides of uranium and thorium has been loaded into commercial light water reactors, converting the (predominantly abundant) thorium-232 by neutron capture into fissile uranium-233 (Thorn *et al.*, 1998). In one case - the Shippingport PWR, modified to maximise the neutron economy, a power run over a 5-year period produced more fissile uranium than was consumed. This important result opened up the thorium resources to fission process using established reactor technology. Reprocessing is required, however, and this technology is more difficult for thorium than for uranium fuel, and has not yet been developed. Very little plutonium or other trans-uranic elements are produced in this fuel cycle, but the technology is not proliferation resistant, since the U-233 can be used to make nuclear weapons. The U-233 will, however, be accompanied by U-232, which generates a hard γ - ray emitting decay product, making it more difficult to handle than weapons grade U-235. But, like U-235, it can readily be denatured.

5.5 Accelerator Hybrids: "Incinerators"

An inevitable end product of all the uranium fission processes described above is a substantial residue of plutonium and other long-lived higher actinides, such as americium and curium. "Incineration" (transmutation to non-radioactive, or short-lived radioactive products) of such material calls for a source of neutrons of energies higher than those usually available in conventional reactors. Partly for this reason there has been discussion of designs of super-fast burner reactors, in which the source of high energy neutrons would be a proton accelerator, working in conjunction perhaps with a thorium reactor, part of whose power output would run the accelerator. However, the viability of this technology has been rejected in recent analyses (NRC, 1995) and no such reactors have yet been developed.

5.6 Fusion Reactors

As already mentioned, and illustrated in figure 1, an alternative to fission in exploiting nuclear energy is the *fusion* of two very light atoms to form a single heavier atom: e.g. two atoms of deuterium (H-2) fusing to form one of helium would, in the process, release some 300 TJ per kilogram of starting mass. The engineering challenge is, however, immense, involving the production, confinement and steady maintenance of a state of matter - a plasma - at a temperature above 100 million Celsius. If, and when, such a process could be achieved on an industrial scale it would constitute a source of nuclear power for which fuel supplies were essentially unlimited and which did not generate fissile material.

6. Resources for Nuclear Fission Power

Nuclear power stations generating a total of some 352 Gigawatts of electricity (GW(e)) are currently operating world wide, with a further 27 GW(e) under construction; they generate about 17 % of the world's electricity consumption and 5 % of its total energy usage (IAEA, 1998). All these power plants consume uranium only; they require about 60 kilotonnes of natural uranium metal per annum to operate, consume about 2 % of this and reject the rest as spent fuel. Around $40 per kg is paid for the raw fuel (usually uranium oxide), which constitutes a few percent of the busbar cost of electricity.

The world resource of uranium (excluding Chile and China), extractable at less than $130 per kg, is quoted as 4.4 Mt (OECD, 1998). Estimates of this resource have been increasing year by year at a rate greater than its consumption for electricity generation: one estimate of the ultimate resource is 20 Mt (Bowie, 1993). Evidently, at the present rate of usage, 20 Mt would last some 300 years. Currently only some $70 M per annum is being spent on exploration.

Thus, the question of uranium supplies becomes a pressing one only if nuclear power is greatly expanded. Taking account of the expected expansion of world electricity demand, and of the commitment by many governments to reduce the burning of fossil fuel, a ten-fold expansion of nuclear power in the next century could be contemplated. Such a scenario would concentrate attention on the potential resource limitation, which could be countered by three of the nuclear fission technologies that have been outlined above: use of recycled uranium and plutonium in MOX fuel; use of fast breeder reactors; and the use of thorium in one or other of the existing reactor types. A useful analysis of the broader context of nuclear and other energy resources has given by the (UK) National Academy (1995)

7. Economics

Whilst there is general consensus as to most of the physical background to nuclear power, its economics may be more contentious: all that it will be possible and appropriate to do here is to enumerate some of the considerations that may need to be taken into account.

- Investment in nuclear power entails long-term commitment: even omitting development time, the total period for building, operating and decommissioning a plant will be of the order of 50 years.
- The cost of nuclear power is heavily weighted by capital costs - i.e. the cost of borrowing, or discount rate - with fuel costs relatively low. It is thus economically suited for supplying base-, rather than marginal-load.
- In line with the above, nuclear power is technically suited to continuous, steady-level running.
- As discussed in the previous section, the availability of high grade uranium ore to supply fuel at around current costs is limited. However, low grade sources (including

seawater: see Chapter 11, by Garwin, in this study) are large and it is unclear how soon costs might rise to the point that breeder reactors or other options become attractive.

- As with other energy sources, realistic assessments of cost must be comprehensive, taking into account decommissioning and environmental components. A real difficulty is how to assess future interest-, and thus discount-rates.
- Any decisions on investment in nuclear power will be heavily influenced by comparisons with other options. But this will also entail considerations such as: reliability and local availability of fuel supplies, and the need for diversity in power sources.
- Good cost efficiency in a nuclear power programme is likely to come from well chosen standardisation of design and large scale implementation, e.g. the French model.

A breakdown of the busbar costs of electricity for a pressurised water reactor and for a coal-fired plant is given in table 2. The outstanding feature is the greater proportion of capital cost in the PWR case; a feature common to all nuclear reactors. A significant enhancement of this capital cost can arise from delays both in construction and commissioning and in public regulatory procedures. Capital cost is a deterrent to investment in nuclear power; however, once built and operating, the electricity costs are rather insensitive to fuel price. Substantial information is accumulating about costs of decommissioning and waste disposal, but uncertainties remain; however, because these expenses need not be incurred for many years, their discounted costs are correspondingly reduced.

	Coal Fired	**PWR**
Capital (incl. Decommissioning)	29	43
Operation and Maintenance	17	20
Fuel (transformation)	-	24
Fuel (raw materials)	54	13
Total	100	100

Tab. 2. Relative Electricity Generating Costs in 2050 AD.
Data from Western European experience (Gouni, 1991).

References

Bowie, S.H.U., in *Nuclear Power Technology*, **2**, pp. 56-75 (ed. Marshall). Oxford, OUP, 1998.

Eyre, B.L. *et al.*, eds., "The fast neutron breeder fission reactor", Phil. Trans. Roy. Soc. A, **331,** 287-453, 1990.

Gouni, L., in *Nuclear Fusion*, Supplement **3** (J. Darvas, *et al.*, eds.), pp. 633-646, 1991.

IAEA, *IAEA Bulletin*, **40**, pp. 17, 41. Vienna, International Atomic Energy Authority, 1998.

ICRP, *Annals of the International Commission on Radiological Protection*, Archival periodical journal, Pergamon Press.

ICRP, *1990 Recommendations of the ICRP*, Ann. ICRP, **21,** (1-3), 1990.

ICRP, *Age dependent doses to members of the public from intake of radionuclides*, Ann. ICRP, **26,** (1), 1996.

National Academy, *Energy and the Environment in the 21st Century*, Report by the National Academy's Policy Advisory Group, London, The Royal Society, 1995.

Novikov, I.I. *et al.*, "Reasons for the RBMK reactor accident at the Chernobyl NPP", The Nuclear Engineer, **36,** 142-146, 1995.

NRC, *Nuclear Wastes: Technologies for Separation and Transmutation*, Report of the US National Research Council, Washington D.C., National Academy Press, 1995.

OECD, *Chernobyl Ten Years On: Radiological and Health Impact*, Paris, OECD Nuclear Energy Agency (http://www.nea fr/html/rp/chernobyl/chernobyl.html), 1995.

Price, T., *Political Electricity: What Future for Nuclear Energy?*, Oxford, OUP, 1990.

Roberts, L.E.J. *et al.*, *Power Generation and the Environment*, Oxford, OUP, 1990.

Thorn, J.D., in *Nuclear Power Technology* **2,** (ed. Marshall), pp. 368-413, Oxford, OUP, 1988.

UNSCEAR, *Periodical reports of the United Nations Scientific Committee on the Effects of Atomic Radiation*, UN publication.

UNSCEAR, *Sources, Effects and Risks of Ionizing Radiation*, New York, UN, 1988.

Wilson, P.D., *The Nuclear Fuel Cycle*, Oxford, OUP, 1996.

Chapter 2 Preventing Climate Change: The Role of Nuclear Energy

by
Steve Fetter*

Introduction[3]

In December 1997, parties to the Framework Convention on Climate Change (FCCC) negotiated the Kyoto protocol, in which the industrialized countries agreed to reduce their emissions of greenhouse gases by 5 percent below 1990 levels, by 2008 to 2012. In the United States the agreement has been attacked from both sides, with environmental groups asserting that much deeper reductions are urgently needed, and opponents claiming that the reductions are unnecessary, would curtail economic growth, or would be unfair or ineffective without similar commitments by developing countries.

Both groups overstate the importance of near-term reductions in emissions. The modest reductions called for by the Kyoto agreement are a sensible first step, but only if they are part of a larger and longer-term strategy. The centerpiece of any strategy to achieve the objective of the Climate Convention is a transformation in world energy supply, in which traditional fossil fuels are replaced by energy sources, that do not emit carbon dioxide. This transformation must begin in earnest in the next 10 to 20 years, and must be largely complete by 2050.

Only five energy sources are capable of providing a substantial fraction of the required carbon-free supply in 2050: fission, solar, "decarbonized" fossil fuels, and, to a lesser extent, biomass and wind. Each of these sources currently has significant technical, economic, and/or environmental handicaps. For example, nuclear fission, which. is the only one of the five that is deployed commercially on a large scale today, suffers from concerns about cost and risks related to accidents, waste disposal, and the spread of nuclear weapons. The most urgent need, therefore, is a broad-based program of energy research and development to attempt to ameliorate these concerns, and thereby ensure that inexpensive and acceptable substitutes will be available worldwide when they are needed.

---

* Steve Fetter is Associate Professor and Director of the Environmental Policy Program at the School of Public Affairs, University of Maryland, College Park, Maryland (US).

[3] This paper is available at http://www.puaf.umd.edu/papers/fetter.htm.

1. The Objective of Emission Controls

The objective of the Climate Convention is to achieve "stabilization of greenhouse gas concentrations in the atmosphere at a level that would prevent dangerous anthropogenic interference with the climate system".[4] The level that would prevent "dangerous interference" is undefined, but the Convention states that stabilization "should be achieved within a time-frame sufficient to allow ecosystems to adapt naturally to climate change, to ensure that food production is not threatened and to enable economic development to proceed in a sustainable manner."

Most studies of climate change focus on the effects of a doubling of the carbon dioxide concentration from the pre-industrial level of about 280 parts per million (ppm). According to the Intergovernmental Panel on Climate Change (IPCC), a doubling would, over the long term, increase the global-average surface air temperature by 1.5 to 4.5 °C, with a best estimate of 2.5 °C.[5] The wide range is due largely to uncertainties about how cloud cover, ocean currents, and vegetation would change as the atmosphere warmed. More important than changes in average global temperature, but even more difficult to predict, are regional changes in seasonal temperature, precipitation, and soil moisture, and in the frequency of extreme events such as storms and drought. In general, average temperature increases in northern continental regions are expected to be twice the global average. Average precipitation is predicted to increase by 5 to 15 percent, but some regions, such as the interior of North America, are expected to become drier in the summer because of even greater increases in evaporation.[6]

Would these changes constitute "dangerous interference" with the climate system? One way to gain insight is to examine past changes in climate. Figure 1 shows, in a schematic way, how the average temperature of the Earth has varied over the last million years. Also shown are estimates of future changes expected in a "business-as-usual" scenario, in which greenhouse gas concentrations reach an equivalent doubling by 2070 and continue to rise thereafter. Several features of this temperature history deserve special attention.

---

[4] United Nations Framework Convention on Climate Change, May 1992, http://www.unfccc.de.

[5] "Technical Summary" in J.T. Houghton, L.G. Meira Filho, B.A. Callander, N. Harris, A. Kattenberg and K. Maskell (eds.), *Climate Change 1995; The Science of Climate Change*, Cambridge University Press, 1996, p. 34.

[6] A. Kattenberg, F. Giorgi, H. Grassl, G.A. Meehl, J.F.B. Mitchell, R.J. Stouffer, T. Tokioka, A.J. Weaver, T.M.L. Wigley, "Climate Models-Projections of Future Climate", in J.T. Houghton, *et al.* (eds.), *Climate Change 1995; The Science of Climate Change*, pp. 291-357; Edward Bryant, *Climate Process and Change*, Cambridge University Press, 1997, p. 134.

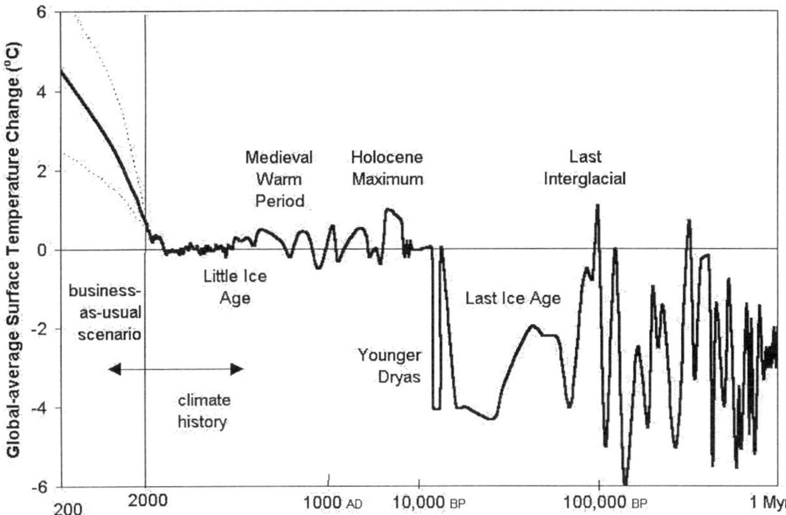

Fig. 1. Global-average surface temperature change over the last million years, and projected change to 2200 under a "business-as-usual" scenario. *Sources: 1 million years BP to 1400 AD: C.K. Folland, T.R. Karl, and K.Y.A. Vinnikov, "Observed Climate Variations and Change", in J.T. Houghton, G.J. Jenkins, and J.J. Ephraums, eds., Climate Change: The IPCC Scientific Assessment (Cambridge: Cambridge University Press, 1990), p. 202; 1400 to 1850 AD: Michael E. Mann, Raymond S. Bradley, and Malcolm K. Hughes, "Global-scale Temperature Patterns and Climate Forcing over the Past Six Centuries", Nature, Vol. 392, 23 April 1998, pp. 779–787; 1850 to 1998 AD: J. Hansen, R. Ruedy, Mki. Sato, and R. Reynolds, "Global Surface Air Temperature", http://www.giss.nasa.gov/data/gistemp; 2000 to 2200 AD: M. Hulme, S.C.B. Raper, and T.M.L. Wigley, "An Integrated Framework to Address Climate Change (ESCAPE) and further Developments of the Global and Regional Climate Models (MAGICC)", Energy Policy, Vol. 23, 1995, pp. 347-355, assuming a total radiative forcing in 1990 of 1.32 W/m^2 and a climate sensitivity of 1.5 to 4.5 C, for the range of IS92 scenarios.*

First, global average temperature has increased by about 0.5 C over the last 70 years, consistent with estimates based on the increase in greenhouse gases during this period.[7] This warming has been accompanied by the retreat of mountain glaciers, a

[7] The observed increase in global-average surface temperature from the mid-19th century to 1990 is 0.45 ± 0.15 °C (N. Nicholls, G.V. Gruza, J. Jouzel, T.R. Karl, L.A. Ogallo, and D.E. Parker, "Observed Climate Variability and Change", in Houghton, *et al.* (eds.), *The Science of Climate Change,* p. 143). The predicted increase from 1765 to 1990 is 0.3 to 0.6 °C. (Results of the model described in M. Hulme, S.C.B. Raper, and T.M.L. Wigley, "An Integrated Framework to Address Climate Change (ESCAPE) and further Developments of the Global and Regional Climate Models (MAGICC)", *Energy Policy*, Vol. 23, 1995, pp. 347-355, assuming a total radiative forcing in 1990 of 1.32 W/m^2 and a climate sensitivity of 1.5 to 4.5 °C.)

1 percent increase in precipitation over land, an increase in cloud cover, and a 10 to 25 cm rise in sea level - all of which are consistent with predictions based on an enhanced greenhouse effect.[8] The last decade was the warmest period since at least the 14th century,[9] and one of the warmest in the last 10,000 years.

Second, average temperature has been relatively stable for the last 10,000 years, with variations of about 1 °C. This period of stable climate coincides with the development of agriculture and human civilization. However, even these relatively small variations in global average temperature were associated with significant changes in regional climate that had important consequences for ecosystems and human societies. For example, 4000 to 6000 years ago, when global average temperature was about 1 C higher than at present, the tropics were wetter and experienced catastrophic floods four to ten times greater than those witnessed today, and temperate latitudes were significantly drier.[10] Between 1100 and 1300 AD, when temperatures in Europe were about 1 C higher than at present, the Vikings colonized Greenland. The subsequent cool period known as the "Little Ice Age," when average temperatures in Europe and China were 0.5 to 1 C lower than at present, was accompanied by violent storms and floods, crop failures, widespread famine, and devastating epidemics.[11]

Third, over the last two million years the climate has oscillated between long ice ages and shorter interglacial periods, with a period of about 100,000 years. During the last ice age, average temperatures and sea levels were about 5 C and 120 meters lower than at present; during the last interglacial period, temperatures and sea levels were about 2 C and 5 meters higher than present. These changes in temperature, which were accompanied by dramatic shifts in the distribution of vegetation, are comparable to that which would accompany a doubling of the carbon dioxide concentration.

Glacial periods are correlated with known variations in the Earth's orbit, which change the amount of summer sunshine received by the poles. These variations in sunshine are too small, by themselves, to account for the observed changes in climate. There must exist strong feedback mechanisms in the climate system - for example, changes in the biosphere or ocean currents - which serve to amplify the warming caused by increased sunshine. The sensitivity of the climate system to past variations in sunshine should make us wary about its sensitivity to changes in the radiation balance caused by increased greenhouse gas concentrations.

Fourth, past shifts in climate sometimes have been very rapid. For example, there were about two dozen instances during the last ice age when temperatures rose or fell by up to 5 C over periods of less than a few decades. As the Earth emerged from the last ice age 13,000 years ago, the climate suddenly returned to ice-age conditions; 1300 years later, a warming in the Arctic of 5 to 10 C occurred over several decades or less, after

---

[8] Nicholls, *et al.*, "Observed Climate Variability and Change", p. 149, 156, 163; R.A. Warrick, C. Le Provost, M.F. Meier, J. Oerlemans, and P.L. Woodworth, "Changes in Sea Level", in Houghton, *et al.* (eds.), *The Science of Climate Change,* p. 366.

[9] Michael E. Mann, Raymond S. Bradley, and Malcolm K. Hughes, "Global-scale Temperature Patterns and Climate Forcing over the Past Six Centuries", *Nature*, Vol. 392, 23 April 1998, pp. 779–787.

[10] Bryant, *Climate Process and Change,* p. 90, 192.

[11] Bryant, *Climate Process and Change,* p. 90-91, 157.

which the current warm climate has prevailed.[12] These rapid shifts in climate may have been caused by a switching on and off of the North Atlantic thermohaline circulation, which today transports huge quantities of heat northward, keeping Europe much warmer than other regions of the same latitude. These episodes alert us to the possibility that rapid, large-scale changes in climate might be triggered if temperatures increase beyond some threshold. Although the threshold, if one exists, is unknown, it might be no greater than the upper range of the temperature increase predicted for a doubling of carbon dioxide.[13] Such an event, if it happened today, would have devastating effects on global agriculture and human civilization.[14]

Another way to gain insight into how much change would be dangerous is to model the effects of climate change on ecosystems, agriculture, and economies. In general, an increase in carbon-dioxide concentrations, and the associated increase in global average temperature and precipitation, should promote plant growth, except in areas where the additional precipitation does not compensate for the increase in evaporation. Under the climate conditions predicted for a doubling of the carbon dioxide concentration, models indicate that present-day vegetation patterns would change over 20 to 40 percent of the world's surface area. Current vegetation boundaries would shift by 300 to 1,000 kilometers, greatly outstripping the ability of most species to migrate naturally.[15] Rising sea levels will also cause wetlands to be lost at a faster rate than new wetlands would be created.

The capacity of human societies to modify agricultural practices in response to changes in climate is much greater than during previous periods of change, particularly in developed countries. One study concluded that, for climate conditions predicted for a doubling of carbon dioxide, total world grain production would decline by up to 5 percent, compared. to what it would have been without climate change.[16] With a greater degree of adaptation (changes in crops and additional irrigation), the study concluded that global harvests could be maintained at no-climate-change levels, but that production in developing countries would nevertheless decline significantly. The effects of increased climate variability or disruptions caused by sudden shifts in climate have not been examined.

Much attention has been given to the economic costs of climate change and of mitigating greenhouse gas emissions. Most studies have included costs associated with

---

[12] Jeffrey P. Severinghaus, Todd Sowers, Edward J. Brook, Richard B. Alley, and Michael L. Bender, "Timing of Abrupt Climate Change at the End of the Younger Dryas Interval from Thermally Fractionated Gases in Polar Ice", *Nature*, Vol. 391, 8 January 1998, pp. 141–146; Scott Lehman, "Sudden End of an Interglacial", *Nature*, Vol. 390, 13 November 1997, pp. 117–119.

[13] Thomas F. Stocker and Andreas Schmittner, "Influence Of CO_2 Emission Rates on the Stability of the Thermohaline Circulation", *Nature*, Vol. 388, 28 August 1997, pp. 862-865.

[14] William H. Calvin, "The Great Climate Flip-flop", *The Atlantic Monthly*, Vol. 281, No. 1, January 1998, pp. 47-64.

[15] T.M. Smith, R. Leemans, and H.H. Shugart, "Sensitivity of Terrestrial Carbon Storage to CO_2-induced Climate Change: Comparison of Four Scenarios based on General Circulation Models", *Climate Change*, Vol. 21, pp. 367-384; and R.A. Monserud, N.M. Tchebakova, and R. Leemans, "Global Vegetation Change Predicted by the Modified Budyko Model", *Climate Change*, Vol. 25, pp. 59-83.

[16] Cynthia Rosenzweig and Martin L. Parry, "Potential Impact of Climate Change on World Food Supply", *Nature*, Vol. 367, 13 January 1994, pp. 133-138.

sea-level rise, forest and fishery losses, and changes in agriculture, energy demand, hurricane damage, and water supply, but have ignored or underestimated impacts that are difficult to monetize, such as the value of ecosystem and species loss, air and water pollution, and human death, illness, discomfort, and aesthetics. As with studies of ecosystem and agricultural impacts, cost studies generally have not considered the effects of possible increases in climate variability or rapid changes in climate.

With these caveats in mind, the expected cost of impacts associated with a 2.5 C average temperature increase is estimated at 1 to 2 percent of gross domestic product (GDP) for developed countries, 2 to 9 percent for developing countries, and about 2 percent for the world as a whole.[17] For some countries, such as low-lying islands, losses could be a much greater percentage of GDP. For comparison, 2 percent of current gross world product (GWP) is over $500 billion per year.

There is, of course, great uncertainty in these estimates. In a poll of 19 experts, best guesses of the cost of a 3 C warming by 2090 ranged from 0 to 20 percent.[18] Half believed that there is at least a 10 percent chance that the cost would be greater than 6 percent of GWP. The average respondent believed that costs would triple if the average temperature increase were 6 C instead of 3 C, and that there would be a 5 percent chance of a 25 percent drop in GWP - the rough equivalent of the Great Depression.

2. Selecting a Stabilization Target

One way to develop a strategy is to construct reasonable scenarios and to ask what we should be doing today if these scenarios were to become reality. We do not know very accurately how climate will change in response to increased greenhouse-gas concentrations, or how natural systems and human societies will be affected by changes in climate. But it is worthwhile to set tentative limits on greenhouse gas concentrations based on the current state of knowledge, trace the implications of such limits for the future of world energy supply, and to ask what we should be doing today to prepare for these changes.

Based on what we know today, it would be very difficult to justify a stabilization target greater than an equivalent doubling of carbon dioxide. Stabilization at this level would result in an eventual increase in average temperature of as much as 5 °C, and 2.5 C over the next century, entailing a significant risk of catastrophe. Even the "best estimate" change in temperature - 2.5 C total and 1.5 C over the next century - would entail significant risk of costly changes in climate, particularly in the northern regions.

The stabilization target can be expressed in terms of the "instantaneous radiative forcing," or the change in the energy balance of the climate system that would result from

[17] D.W. Pearce, W.R. Cline, A.N. Achanta, S. Fankhauser, R.K. Pachauri, R.S.J. Tol, and P. Vellinga, "The Social Costs of Climate Change: Greenhouse Damage and the Benefits of Control", in James P. Bruce, Hoesung Lee, and Erik F. Haites (eds.), *Climate Change 1995; Economic and Social Dimensions of Climate Change*, Cambridge University Press, 1996, pp. 203-205.

[18] William D. Nordhaus, "Expert Opinion on Climate Change", *American Scientist*, Vol. 82, Jan/Feb 1994, pp. 45-51.

an instantaneous change in greenhouse-gas concentrations. A doubling of carbon dioxide produces a radiative forcing of 4.4 watts per square meter (W/m^2); an "equivalent doubling" is any set of greenhouse-gas concentrations that produce a combined forcing of 4.4 W/m^2. Over the last 150 years, deforestation and the burning of fossil fuels have increased the concentration of carbon dioxide from about 280 ppm to 364 ppm. The total radiative forcing, including contributions from other long-lived greenhouse gases, is 2.6 W/m^2, which is equivalent to a carbon-dioxide concentration of about 420 ppm.[19] Thus, we already are halfway toward an equivalent doubling of carbon dioxide.

3. Limits on Fossil-fuel Emissions

To translate a stabilization target into a limit on global emissions of carbon dioxide from the burning of fossil fuels, we must subtract the contributions of greenhouse gases other than carbon dioxide, use carbon-cycle models to determine rates of emission that would lead to stabilization at the desired level, and account for carbon dioxide emissions from other sources, such as land-use changes and cement manufacture.

3.1 Other greenhouse gases

Carbon dioxide is the most important greenhouse gas and is more amenable to monitoring and control than other gases, but we must also consider emissions of methane, nitrous oxide, and halocarbons, which are long-lived greenhouse gases. Increased concentrations of these gases currently are responsible for a radiative forcing of 0.9 W/m^2, equivalent to an additional 60 ppm of carbon dioxide. The long-term effect of ozone and various aerosols can be ignored in this context.[20]

Anthropogenic emissions of methane and nitrous oxide are due primarily to agricultural and waste disposal activities. Strategies exist for reducing emissions from most sources, but the practical potential is limited. For example, the largest source of methane emissions - domestic livestock - could be reduced by up to 40 percent through improvements in feeding and manure management[21] but such reductions likely would be

---

[19] Assumes 1997 concentrations and radiative forcings of 363 ppm and 1.64 W/m^2 for carbon dioxide, 1.76 ppm and 0.49 W/m^2 for methane, 0.315 ppm and 0.16 W/m^2 for nitrous oxide, and a forcing of 0.28 W/m^2 for various halocarbons. The equivalent carbon dioxide concentration, C_{eq}, is the CO_2 concentration that would produce a radiative forcing equal to that from all greenhouse gases (in this case, 1.64 + 0.49 + 0.16 + 0.28 = 2.57 W/m^2); $C_{eq} = C_0 \cdot e^{\Delta F/6.3} = (280)e^{(2.57/6.3)} = 421$ ppm.

[20] The influence of ozone and aerosols on climate is highly uncertain. Because their residence times are on the order of days, any effect on climate will be regional, not global. In any case, reductions in fossil-fuel burning will result in proportional decreases in the concentrations of ozone and aerosols, and efforts to control air pollution and acid deposition will lead to reductions in ozone and aerosol concentrations independent of efforts to limit fossil fuel burning, particularly as pollution-control technologies advance and diffuse to developing countries.

[21] Vernon Cole, "Agricultural Options for Mitigation of Greenhouse Gas Emissions", in Robert T. Watson, Marufu C. Zinyowera, and Richard H. Moss (eds.), *Climate Change 1995; Impacts, Adaptations and Mitigation of Climate Change: Scientific-Technical Analyses*, Cambridge University Press, 1996, p. 764.

more than offset by an increase in the number of animals. Similar arguments can be made for most other anthropogenic sources of methane and nitrous oxide. Moreover, natural emissions of these gases may increase as a result of climate change. Thus, even if aggressive efforts are made to limit emissions of methane and nitrous oxide, significant reductions in long-term, total emissions are not likely. If rates of emission remain constant at today's levels, the combined radiative forcing of these two gases would increase from 0.65 W/m² to about 1.0 W/m².[22]

Halocarbons also contribute to greenhouse warming. Although the Montreal Protocol and its Amendments will lead to a phase-out of substances containing chlorine and bromine, their residence times are so large that significant concentrations will remain in the atmosphere for over a hundred years. In addition, many CFC-substitutes, as well as a number of other unregulated substances, are greenhouse gases. Today, the forcing from halocarbons and other trace gases is about 0.28 W/m²; long-term values might be somewhat lower or higher.

For stabilization at an equivalent doubling, gases other than carbon dioxide are likely to contribute a radiative forcing of 1.3 ± 0.4 W/m². Carbon dioxide would then be limited to a forcing of 3.1 ± 0.4 W/m² and a concentration of about 460 ± 30 ppm. At current growth rates, such concentrations would be attained in 40 to 80 years.

3.2 Carbon emissions

Carbon dioxide emitted into the atmosphere is gradually absorbed by the oceans and by plants. Carbon-cycle models, which simulate these processes, can be used to estimate the rates of emission that would result in stabilization of the carbon dioxide concentration at a given level. Figure 2 shows the rate of emission over the next 150 years for stabilization at 450 to 500 ppm. The uncertainty in the emission pathway, which is mostly due to uncertainties about the fertilization of plant growth, is indicated for the 450 ppm case by the two narrow dotted lines. Also shown are emissions for a more gradual approach to 450 ppm and for a more rapid approach to 500 ppm. Two features of this figure are worthy of attention.

First, carbon-dioxide emissions must peak no later than 2020. This conclusion is insensitive to assumptions about other greenhouse gases, the rate at which stabilization is achieved, or model parameters. After peaking, carbon-dioxide emissions must decline to levels below the current rate of emission (about 7.5 PgC/y)[23] by 2050.

---

[22] If rates of emission remain constant at today's levels, concentrations would rise from 1.76 to about 1.90 ppm for methane and from 0.315 to about 0.41 ppm for nitrous oxide. Radiative forcing would increase from 0.49 to 0.53 W/m² for methane and from 0.16 to 0.50 W/m² for nitrous oxide. Results of the MAGICC computer model with methane and nitrous oxide emissions (natural plus anthropogenic) set at 535 and 13.6 Mt/y. T.M.L. Wigley, S.C.B. Raper, M. Salmon, and M. Hulme, *MAGICC: Mode for the Assessment of Greenhouse-gas Induced Climate Change,* Norwich (UK), Climate Research Unit, University of East Anglia, April 1997.

[23] 1 Pg (petagram) = 10^{15} g = 10^{9} tonnes.

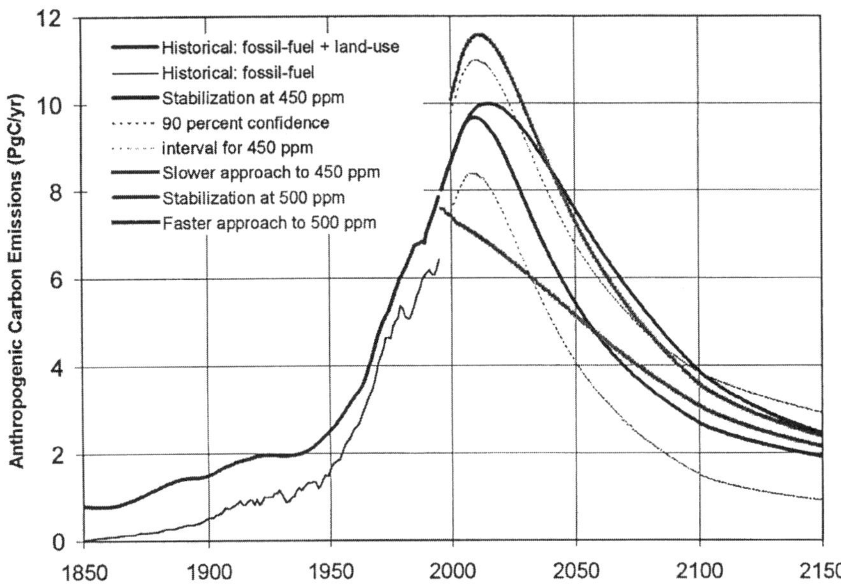

Fig. 2. Historical emissions of carbon from fossil-fuel burning and land-use changes, and emission pathways that stabilize carbon dioxide concentrations at 450 and 500 ppm in the period 2100 to 2150.[24] *Source: Author's calculations based on results from the model described in T.M.L. Wigley, "Balancing the Carbon Budget: Implications for projections of Future Carbon Dioxide Concentration Changes", Tellus, Vol. 45B, pp. 405-425.*

Second, the stabilized concentration of carbon dioxide is determined primarily by the rate of emission in the second half of the next century. A slower approach to stabilization would require immediate reductions in emissions, but would permit only slightly higher emissions over the long term. Conversely, a more rapid approach to stabilization would allow much higher emissions in the near term at the expense of slightly lower emissions over the long term. The total amount of carbon dioxide that can be emitted over the next 100 to 150 years is larger for a more-rapid approach to stabilization because near-term carbon emissions will largely be absorbed by the oceans and the biosphere by the time stabilization is achieved. In other words, stabilization at an equivalent doubling can be achieved even if emissions increase substantially over the next 10 to 20 years, as long as emissions are reduced below the current level by 2050.

[24] The original of this figure was depicted in colours. The two solid lines, from 1850 to a little before 2000, are the historical fossil fuel and land-use, and fossil fuel only, carbon emission lines. Of the two (middle) solid curves starting at the same point, in the year 2000, the one with the 90 % confidence level dotted lines "symmetrically" around it is the 450 ppm stabilization line, while the other ("upper") one is the 500 ppm stabilization line. The solid curve starting at the highest point, in the year 2000, is the more gradual approach to 450 ppm. The curve starting at the lowest point, in the year 2000, is the more rapid approach to 500 ppm.

Thus, the stabilization target can be translated into a target for total carbon emissions in 2050. Near-term reductions in emissions are important primarily insofar as they help achieve the target in 2050. In general, it is probably better to invest money in future reductions (via energy research and development) than to pay for costly reductions today.[25]

3.3 Non-fossil-fuel carbon emissions

Although anthropogenic carbon-dioxide emissions are due mostly to fossil-fuel burning, deforestation, climate feedbacks, and cement manufacture may also make significant contributions. During the 1980s, net deforestation released 1.1 ± 0.7 PgC/y.[26] Future emissions are a matter of speculation. Reference scenarios developed by the IPCC and others assume rates ranging from 0 to 2 PgC/y in 2050.[27] On the other hand, scenarios that assume strong policy efforts to slow tropical deforestation and implement reforestation. programs result in a net uptake of carbon of 0.5 to 2.2 PgC/y in 2050.[28] In addition, climate change itself might cause large releases of carbon during the next century if mature forests die before they are replaced by new forests, if higher temperatures promote the decay of dead organic materials at high latitudes, or if drier conditions increase the frequency of forest fire. It is estimated that such processes could release up to 240 PgC over the next century, at rates of up to 3 PgC/y.[29] On the other hand, a warmer, wetter climate might result in the expansion of tropical and boreal forests, leading to a net absorption of up to 100 PgC over several centuries.[30] Finally, carbon dioxide is released during the manufacture of cement. In 1995, cement manufacture released 0.2 PgC; by 2050, this could be expected to increase to 0.5 PgC/y. I estimate that land-use changes, climate feedbacks, and cement production will result in a net release of 60 ± 50 PgC during the next 50 years (1 ± 1 PgC in 2050).

---

[25] A possible exception is if climate change is highly sensitive to the rate of increase of greenhouse gases, as well as the ultimate stabilization level. In that case, near-term reductions and a more gradual approach to stabilization might make sense.

[26] D. Schimel, *et al.*, "Radiative Forcing of Climate", in Houghton, *et al.* (eds.), *Climate Change 1995; The Science of Climate Change*, pp. 78-79.

[27] J. Alacamo, A. Bowman, J. Edmonds, A. Grubler, T. Morita, and A. Sugandhy, "An Evaluation of the IPCC IS92 Emission Scenarios", in J.T. Houghton, L.G. Meira Filho, J. Bruce, Hoesung Lee, B.A. Callander, E. Haites, N. Harris and K. Maskell (eds.), *Climate Change 1994; Radiative Forcing of Climate Change and An Evaluation of the IPCC IS92 Emission Scenarios*, Cambridge University Press, 1995, pp. 284-286.

[28] Alacamo, *et al.*, "An Evaluation of the IPCC IS92 Emission Scenarios", p. 286; Sandra Brown, "Management of Forests for Mitigation of Greenhouse Gas Emissions", in Watson, Zinyowera, and Moss (eds.), *Climate Change 1995; Impacts, Adaptations and Mitigation of Climate Change*, p. 775.

[29] Miko U.F. Kirschbaum and Andreas Fischlin, "Climate Change Impacts on Forests", in Watson, Zinyowera and Moss (eds.), *Climate Change 1995; Impacts, Adaptations and Mitigation of Climate Change*, p. 104.

[30] See citations in Melillo, *et al.*, "Terrestrial Biotic Responses", p. 466.

3.4 Fossil-fuel emissions

Emissions of carbon from fossil-fuel burning have risen steadily over the last half century, from about 1.4 PgC/y in 1945 to 6.2 PgC/y in 1995 - an average growth rate of 3 percent per year.[31] Including other sources of carbon, total anthropogenic emissions were 7.5 ± 0.9 PgC in 1995.

In order to stabilize greenhouse gas concentrations at an equivalent doubling, fossil-fuel carbon emissions must be limited to 5 ± 2.5 PgC/y in 2050. Given projected population increases, this will be equal to a global average of about 0.5 tC/y per capita in 2050 - a level of fossil-fuel emissions that has not been seen since the end of World War II. Assuming that 1 EJ of fossil energy releases about 18 TgC, fossil-fuel energy consumption would be limited to 280 ± 140 EJ/y (9 ± 4.5 TW) in 2050, compared with 330 EJ/y (10 TW) in 1995.

4. Carbon-free Energy Supply

The demand for energy will grow substantially over the next century, driven by increases in both population and per-capita consumption in developing countries. Figure 3 shows several scenarios of future energy consumption. Except for the "WEC C" scenario, they assume no special policies to decrease energy consumption or carbon emissions, but they do take into account expected improvements in energy efficiency and price increases caused by the depletion of oil and gas resources. Estimates of world primary energy consumption range from 600 to nearly 1300 EJ/y (19 to 42 TW) in 2050. The wide range is due to uncertainties in population forecasts, in future rates of regional economic growth, and in the rate at which energy efficiency is improved.

By subtracting limits on fossil-fuel supply from total energy demand, we derive requirements for carbon-free energy supply.[32] These are given in table 1 for stabilization at an equivalent doubling of carbon dioxide. Note that the supply of energy from sources that do not emit carbon must grow from 53 EJ/y (1.7 TW) in 1995 to 500-1000 EJ/y (16 to 32 TW) by 2050 - an average growth rate of 4 to 5.5 percent per year.

---

[31] Gregg Marland, Tom Boden and Bob Andres, *Revised Global CO_2 Emissions from Fossil-fuel Burning, Cement Manufacture, and Gas Flaring: 1751-1995*, NDP-030/R8, Oak Ridge National Laboratory, 9 January 1998, http://cdiac.esd.ornl.gov/ftp/ndp030/global95.ems.

[32] The difference between total demand and fossil supply could be narrowed by reductions in demand caused by market interventions, such as carbon taxes, which are beyond the scope of this paper.

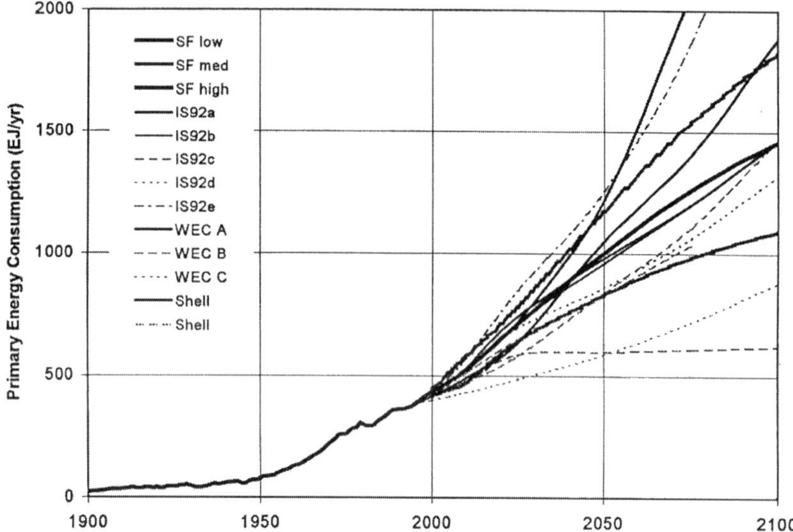

Fig. 3. Scenarios of future world commercial primary energy consumption by Fetter (SF), the Intergovernmental Panel on Climate Change (IS92), the World Energy Council (WEC), and Shell Oil. *Sources: Steve Fetter, Climate Change and the Transformation in World Energy Supply, Stanford, CA, Center for International Security and Cooperation, 1999; J. Leggett, W.L. Pepper, and R.J. Swart, "Emission Scenarios for IPCC: An Update", in J.T. Houghton, B.A. Callander and S.K. Varney (eds.), Climate Change 1992: The Supplementary Report to the IPCC Scientific Assessment, Cambridge, Cambridge University Press, 1992; World Energy Council and International Institute of Applied Systems Analysis, Global Energy Perspectives to 2050 and Beyond, London, WEC, 1995; and Shell International Ltd., The Evolution of the World's Energy Systems, London, Shell International, 1996.*

The implications of this scenario for world energy supply are profound. Today, fossil fuels supply 86 percent of commercial energy supply. If greenhouse gases are to be stabilized at an equivalent doubling, traditional fossil fuels can supply no more energy in 2050 than they supply today, even while total energy use doubles or triples. Carbon-free sources must grow from 14 percent of total commercial supply to 60-80 percent of total supply in 2050.

The transition to carbon-free sources will be the third transformation in world energy supply. The first shift, from firewood to coal, took place from 1850 to 1900. The second shift, from coal to oil and gas, occurred from 1925 to 1975. In these first two shifts, it took 50 years for the emerging source to go from 10 to 60 percent of total supply. The third major shift, from fossil fuels to carbon-free sources, will occur from 2000 to 2050 - if we decide to take seriously the goal of preventing dangerous interference with the climate system.

	Commercial Primary Energy Supply (EJ_p/y)[33]			Growth of non-CO_2 Supply (%/y)	
Year	**Total**	**Limit on Fossil Fuels**	**Non-CO_2 Supply**	**since 1995**	**previous 25 years**
1995	382	329	52.9	2.1	5.7
2025	710 ± 130	430 ± 130	280 ± 180	5.7 (+1.8/–3.6)	5.7
2050	1000 ± 220	270 ± 140	730 ± 260	4.9 (+0.6/-0.9)	3.9
2075	1250 ± 600	210 ± 130	1040 ± 610	3.8 (+0.6/-1.1)	1.4
2100	1450 (+1300/-700)	150 ± 90	1300 (+1300/-710)	3.1 ± 0.6	1.0
2150	1700 (+2000/-800)	50 ± 70	1650 (+2000/-800)	2.2 ± 0.5	1.0

Tab. 1. World commercial primary energy supply, traditional fossil supply for stabilization at an equivalent doubling of carbon dioxide, required carbon-free supply, and average growth rate of carbon-free supply. *Sources: Figures 2-3 and author's calculations.*

Today, only two carbon-free sources - hydropower and nuclear fission - produce significant amounts of energy, with each accounting for about 26 EJ/y (0.82 TW) or 7 percent of commercial primary energy in 1995. Traditional biomass fuels provide 50 to 60 EJ/y (1.6 to 1.9 TW), but much of this is supplied by fuel-wood that is harvested in a unsustainable manner, resulting in a net release of carbon dioxide. Non-fossil energy supply has been growing recently at only about 2 percent per year - much less than the 5 percent-per-year rate needed to stabilize greenhouse-gas concentrations at an equivalent doubling. We will need at least 500 EJ_p/y (16 TW) of carbon-free energy by 2050. Where will this energy come from?

Only five sources are capable of providing a substantial fraction of this non-carbon supply: solar, fission, "decarbonized" fossil fuels, and, to a lesser extent, biomass and wind. Other potential sources are either too limited (hydro, tidal power, and hot-water geothermal), too expensive (ocean thermal and wave energy), or too immature (fusion and hot-rock geothermal) to make a substantial contribution by 2050. Each of the five major alternatives currently has significant technical, economic, and/or

[33] 1 EJ/y = 31.69 GW.

environmental handicaps. Solar is benign but expensive, and would require massive energy storage or intercontinental transmission. Fission can produce electricity at competitive prices today, but suffers from public-acceptance problems related to the risks of accidents, waste disposal, and the spread of nuclear weapons. Coal is cheap and abundant, but the cost and environmental impact of capturing, transporting, and disposing of the carbon dioxide could be high. Biomass has the potential to supply low-cost portable fuels, but energy crops could compete with food production and the preservation of natural ecosystems. Wind is economically competitive in certain areas, but attractive sites are limited.

The most pressing need, therefore, is research and development aimed at reducing the liabilities of the major alternatives. Last year, the US government spent a little more than $1 billion on energy R&D, compared with the $500 billion spent on energy in the United States ($60 billion of which went for imported oil). Total energy R&D - private as well as public - amounted to less than 1 percent of energy expenditures, compared with an average of 3.5 percent for all US industries.

In the past, it has taken about 20 years to realize significant commercial benefits from energy research and development. To prepare for - and profit from - the transformation in energy supply that must begin in earnest by 2015, we must do the R&D today. Our options are limited. We are not smart enough to pick sure winners, and the stakes are too high to rule out any major alternative. We need a balanced R&D program that includes substantial investments in all the sources mentioned above, including nuclear fission.

5. The Potential Role of Fission

Of the carbon-free sources that could make a major contribution to energy supply in 2050, fission is the only source that is deployed commercially on a significant scale today. In 1996, fission supplied 19 percent of world electricity and over 6 percent of commercial primary energy. Near-term prospects for nuclear power are not very favorable. Forecasts range from a substantial decrease to a slight increase in installed capacity over the next 20 years, with fission's share of total world electricity production falling to less than 10 percent by 2020.[34] This is due a combination of factors: the availability of cheaper alternatives, the retirement of older plants, and public opposition to nuclear power in many countries due to concerns about accident and waste-disposal

[34] The OECD projects total capacity to grow from 353 GW_e in 1996 to 400–500 GW_e by 2015, for an average growth rate of 0.6 to 1.9 percent per year. (NEA and IAEA, *Uranium 1997*, p. 60.) EIA projections range from 170 to 420 GW_e in 2020, compared to 351 GW_e in 1996. In the EIA reference case, nuclear generates 2020 of 23,150 TWh in 2020 (8.7 percent). In the low economic growth scenario, nuclear generates 1750 of 18,360 TWh (9.5 percent); in the high economic growth scenario, nuclear generates 2360 of 27,190 TWh (8.7 percent). (Energy Information Administration, *International Energy Outlook 1998*, Washington DC, Department of Energy, p. 89.)

risks and potential links to the spread of nuclear weapons.[35] The only region expected to experience significant growth in the near future is East Asia.

Over the next 50 to 100 years, fission could be expanded to provide half of the world's electric power and a third of the carbon-free[36] supply required to stabilize greenhouse gas concentrations at an equivalent doubling. This is unlikely to happen, however, unless concerns about accidents, waste disposal, and proliferation are resolved.

Prevailing attitudes have conspired to inhibit innovative thinking about these concerns. Most people in the nuclear energy community regard such concerns as political rather than technical in nature. In their view, current reactor designs are very safe, waste-disposal risks are infinitesimal, proliferation risks are purely theoretical, and costs have been inflated (in the United States, at least) by unjustified licensing delays. Conversely, most people in the anti-nuclear community believe that the liabilities of nuclear energy are so great and intractable that no amount of R&D could solve them. In their view, fission is simply "beyond the pale," and government-sponsored research on fission would only divert resources from renewables and prop up an industry that otherwise is headed toward extinction. In the United States, federal funding for fission-energy R&D has declined from nearly \$2 billion in FY78 (Fiscal Year 1978) to a mere \$46 million in FY98, with no funds allocated for new reactor concepts. Industry spending has also declined greatly, reinforced by the movement toward utility deregulation.

This may be changing. In a recent report on US energy research and development, the President's Committee of Advisors on Science and Technology argued that "given the desirability of stabilizing and reducing greenhouse gas emissions, it is important to establish fission energy as a widely viable and expandable option if this is at all possible. A properly focused R&D effort to address the problems of nuclear fission power - economics, safety, waste, proliferation - is therefore appropriate."[37] The key recommendation is the creation of a Nuclear Energy Research Initiative, funded initially at \$50 million per year and increasing over five years to \$100 million per year, to fund R&D on safer and lower-cost reactor designs, new waste-disposal techniques, and proliferation-resistant fuel cycles.

The focus of the proposed program is perfect, but the scale of the effort may be too modest. For comparison, the recommended funding for renewables - mostly biomass, solar, and wind - rises from \$410 to 570 million per year over the five-year period.[38]

---

[35] For a review of the situation in the United States, see Mark Gielecki and James G. Hewlett, "Commercial Nuclear Electric Power in the United States: Problems and Prospects", *Monthly Energy Review*, Washington DC, Energy Information Administration, August 1994.

[36] In scenarios developed by the IAEA and the WEC, nuclear contributes up to 2000 GWe or 150 EJ/y of primary energy by 2050, and 6000 GWe or 450 EJ/y by 2100. See International Atomic Energy Agency, *Nuclear Power: An Overview in the Context of Alleviating Greenhouse Gas Emissions*, IAEA-TECDOC-793, IAEA, Vienna, 1995; World Energy Council and International Institute of Applied Systems Analysis, *Global Energy Perspectives to 2050 and Beyond*, WEC, London, 1995.

[37] President's Committee of Advisors on Science and Technology, Panel on Energy Research and Development, *Report to the President on Federal Energy Research and Development for the Challenges of the Twenty-first Century*, Washington DC, Office of Science and Technology Policy, November 1997, p. ES-19.

[38] PCAST, *Report to the President*, p. ES-33. Also includes geothermal, hydro, and storage for intermittent sources.

Moreover, the Panel recommended that funding for fusion energy - a source which almost certainly will not make a significant contribution to energy supply before 2050 - be increased from $250 to 320 million per year. As another point of comparison, the US government spent about $6 billion, in addition to the billions spent by industry, to help develop the light-water reactor.[39] A serious effort to reinvent fission energy probably would require government support at a rate of several hundred million dollars per year for at least a decade.

What types of fission R&D should be supported? First, R&D is needed on reactor designs that are immune to operator error or equipment failures. Current designs are safe if they are built and operated properly, and advanced versions of these designs are even safer. Unfortunately, examples of poor management of nuclear plants abound.

The goal should be to build reactors that cannot produce off-site fatalities, regardless of what happens inside the plant. The Westinghouse AP 600 reactor, which is nearing design certification, might meet that standard. There should be room in an expanded energy R&D program to support industry-government partnerships on additional advanced designs, such as the Simplified Boiling-water Reactor, the High-temperature Gas Reactor (HTGR), or the Safe Integral Reactor. The concept of small, factory-built modular reactors with lifetime cores is especially interesting.

There is no reason to fund research on breeder reactors for at least the next thirty years. Light-water reactors (LWRs) operating on a once-through fuel cycle could sustain high growth in nuclear electricity production for at least 50 years with conventional uranium resources.[40] Breeder reactors will be economical only if the price of uranium becomes so high that their increased efficiency of uranium use compensates for their higher capital cost, but this is a long way off. Exploration, which has virtually ceased over the last 20 years because of low uranium prices, would undoubtedly uncover substantial additional resources if prices rose significantly. It now appears likely that it will be possible to extract uranium from seawater for less than $250 per kilogram, in which case breeder reactors may never be necessary or economical (See Chapter 11, by Garwin). Given the costs, technical difficulties, proliferation risks, and public-acceptance problems of plutonium use, it would be foolish to unnecessarily tie the expansion of fission over the next 50 or so years to breeder reactors or reprocessing.

Second, governments should support R&D on alternative fuel-cycle concepts designed to minimize proliferation risks in a world with many more reactors, and with reactors in many more countries. This could include novel reactor concepts, such as lifetime cores; new reprocessing techniques that do not involve the separation of pure plutonium; fuel cycles that minimize the production of high-quality plutonium, such as the thorium fuel cycle; the indefinite use of seawater uranium on a once-through fuel cycle; and institutional solutions, such as the consolidation or international control of

---

[39] PCAST, *Report to the President*, p. 5-6. Prior to 1979, the federal government spent about $1.4 billion on light-water-reactor R&D, which is at least $5 billion in 1997 dollars. About $0.8 billion was spent from 1979 to 1997.

[40] For example, in the high-growth scenarios of the IAEA and WEC cited above, installed capacity grows to 1500 to 1900 GWe in 2050, at which point cumulative uranium consumption would be 6 to 9 MtU. Including the lifetime fuel requirements of all reactors then in existence would raise this to 11 to 16 MtU. (Author's estimate.)

facilities that handle plutonium fuels.

Third, governments should support R&D on alternative waste disposal concepts. Today, R&D is limited to a single concept - deep geologic disposal - and, in the United States, to a single site - Yucca Mountain. If current waste-disposal concepts experience significant technical or political setbacks, fission is unlikely to expand substantially. Alternatives to Yucca Mountain should be developed; both short-term alternatives, such as interim storage, as well as long-term alternatives, including disposal in granite and in the deep sea bed.

Conclusion

Meeting the objective of the Framework Convention on Climate Change - to prevent dangerous interference with the climate system - will require a fundamental transformation in the nature of world energy supply beginning in the next 10 to 20 years. Over the next 50 years, the supply of energy by sources that do not emit carbon dioxide must increase ten-fold, from 14 percent to over 60 percent of total supply. All of the possible carbon-free sources have serious economic or environmental drawbacks that must be resolved if they are to play a major role in future energy supply. In the case of fission, we must begin an energetic R&D program to address concerns about accident, waste-disposal, and proliferation risks.

Chapter 3 World Energy and Climate in the Next Century

by
Douglas R.O. Morrison*

Introduction[41]

The world's population has been growing explosively during the last hundred years, tripling from 1.5 billion in 1890 to 5.3 billion in 1990. Even though the rate of increase is currently falling, United Nations (UN) estimates predict a total world population of about 11 to 11.5 billion in 2100.[42]

Energy consumption has a close relationship to this population. The total energy consumption of the world in 1996 was 8380 MTOE (million tons of oil equivalent) or about 400 EJ/y (12.7 TW).[43] The energy consumption has been growing faster than the population. How will the energy needs of the world population be met in the next century, while preserving our environment and avoiding a climate change?

1. Energy Consumption and the Environment

The fuels used, or considered for use, to provide the energy required, can roughly be grouped in three categories:

- *Fossil fuels*, such as coal, oil and natural gas, which currently supply some 90 % of the world's energy;
- *Solar energy*, such as photosynthesis of biomass, hydroelectricity, photovoltaics, wind and waves;
- *Conversion of nuclear mass to energy*, such as in fission of heavy elements like uranium, plutonium and thorium, in fusion of light elements like hydrogen isotopes, and in geothermal power.

It is interesting to plot the ratio $f/(1-f)$ of the fraction f of the total energy supplied by a given source to the fraction $(1-f)$ supplied by all the other sources, against the time (in years), as has been done in figure 1. This figure shows that, historically, each source has passed through a peak, which occurred in about 1920 for coal, and in 1980 for

---

* Douglas Morrison is an Honorary Staff Member at CERN.

[41] This is a summary, prepared by the editors, of the draft text under the same title, presented by Douglas Morrison at the Pugwash Workshop. The complete text may be obtained from the author.

[42] For all references to figures, pictures, quotes and other information presented here, one is referred to the original long version of this chapter.

[43] Please refer to the Appendix for definitions and equivalents of the units used.

oil. Each source then declined in relative importance. Currently, natural gas and nuclear energy have not yet reached their peak, and are still rising. The shape of the curves appear fairly similar from one fuel to the next, and suggests a basis for predicting future trends.

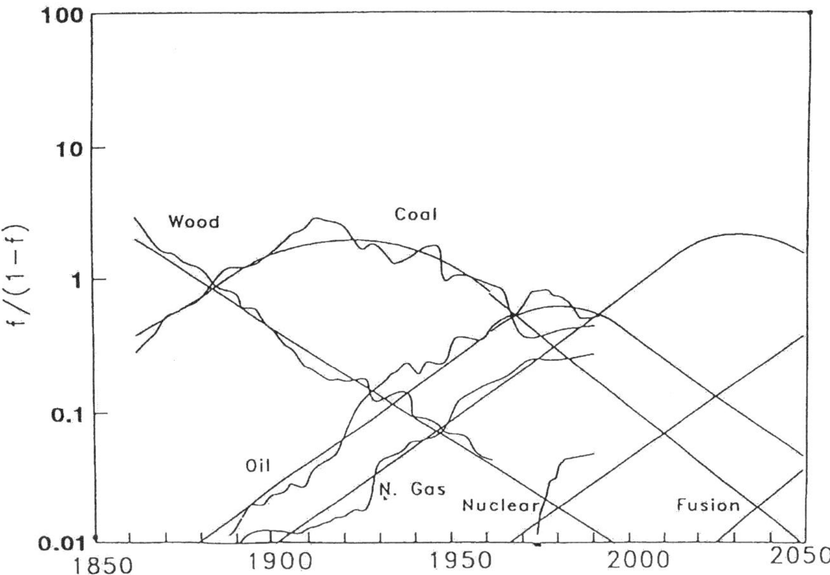

Fig. 1. The fraction *f* of a given energy source with respect to total world energy sources, expressed as $f/(1-f)$. *Source: Morrison, November 1998, extended version of this article.*

Currently, 75 % of the world's energy is consumed by the industrialised countries and only 25 % by the developing countries. Energy consumption of the former has a tendency towards stabilisation, reflecting a rough balance between an actual increase in demand and a concurrent improvement in efficiency of energy use related to advances in technology and improved living standards. For developing countries, however, an increase in energy consumption *per capita* is a *conditio sine qua non*, if they wish - as appears desirable - to pursue their development.

Should the developing countries attain the current level of energy consumption *per capita* of Western Europe, by around the year 2100, the world's total consumption would have increased about six-fold. The prediction of such a future world energy consumption is demonstrated in figure 2. The figure of a six-fold increase in 100 years is, obviously, only approximate, but some such effect appears unavoidable: it represents a phenomenon which *must* occur, as everyone wishes a reasonable standard of living.[44]

[44] In the words of Cohen: "Wealth is health, poverty kills". For the corresponding reference, see the original version of this text.

This effect is the central message of this chapter: there exists a problem and it must get bigger with time as living standards rise as the developing countries attain prosperity and health.

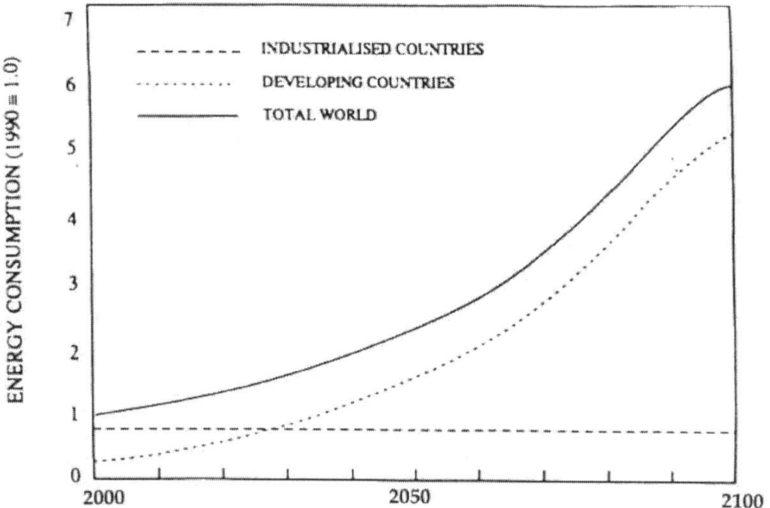

Fig. 2. Prediction of world energy consumption, for industrialised countries (dashed line), for developing countries (dotted line), and the total (broad line). *Source: Morrison, November 1998, extended version of this article.*

Which source will be available to cover these future energy needs? Oil and gas are expected to be in limited supply. Coal production is on the increase in developing countries (China, for instance, mines about 30 % of the world's production of coal), but problems have already been encountered: increased pollution is probably the major one, through the emission of particulate matter, sulphur and nitrogen oxides, but also through the greenhouse effect tied to carbon dioxide emissions. World emissions are currently at the level of 28 million tons per year of sulphur, 45 million tons per year of combined nitrogen and 6 *billion* tons per year of carbon (as CO_2), split about equally between industrialised and developing countries. In this regard, the so-called "poverty trap" plays an important role: developing countries are burning coal in an inefficient and environmentally unfriendly way, because they cannot afford to build the efficient and clean plants needed to solve the problem.

Nuclear power does not create this sort of pollution. It meets, however, with strong opposition, notably from the public opinion, for different reasons:

(a) Worries about the possibility of escaping radiation, as well as doubts about the safety of nuclear reactors, fed by the Three-Mile Island accident and the (more severe) Chernobyl accident,

(b) Problems about waste disposal, both during operation and after decommissioning of nuclear reactors,

(c) The possibility of nuclear proliferation, by the production of nuclear weapons materials in power plants. It would take only about 4 kg of plutonium, or 15 kg of highly enriched uranium, to manufacture one Hiroshima-sized bomb.

Clearly, the risks entailed by the use of each energy source must be weighed and compared. A recent study of the effects of particulate matter on health has shown that if the world consumption of coal in the year 2020 could be brought down from a "business as usual" estimated level of 10.7 billion tons to a more modest level of 8.8 billion tons, through economies and increases in efficiency, *eight million deaths could be avoided*, 80 % of which would be in developing countries.

Fuel reserves are another matter of concern. It has been stated that oil reserves would run out in 40 years, gas reserves in 60 years and coal reserves in about 400 years. These figures must be used with caution, but the inescapable conclusion is that the reserves will run out some time, and that the extraction costs will rise in the meanwhile. Uranium reserves could last much longer. With breeding, fission could last about a thousand times longer than burning coal. It has also been suggested that if the acceptable price for uranium were increased by only a few times, reserves lasting 20 000 years would become available (see also Chapter 11, by Garwin). If fusion ever becomes practicable, there would even be no end of this source in sight.

A first step to deal with the energy problem should be to improve overall energy conservation and efficiency. Different measures have been suggested to promote this idea, including a tax on the use of energy. Carbon-based fuels, however, raise a problem by themselves: the greenhouse effect.

Gases such as CO_2 (carbon dioxide) and CH_4 (methane) in the atmosphere allow much of the sun's radiation to pass through and reach the Earth's surface, but they absorb much of the long wavelength radiation, which is radiated back from our planet's surface. As CO_2 concentrations in the terrestrial atmosphere increase, e.g. by anthropogenic carbon dioxide emissions, this phenomenon could increase the Earth's average surface temperature and bring about overall changes in climate, including the melting of glaciers and polar ice-caps and a subsequent rise of the sea level. This view has not met with general agreement, but the scientific arguments appear to have a definite weight.

The concentration of CO_2 in the atmosphere has been rising at about 1 ppm(v) (parts per million, in volume) per year over the past century. It increased from values varying between extremes of 180 and 260 ppm(v) during the last 100 000 years, to about 350 ppm(v) in 1990. Meanwhile, methane concentrations have gone up from an average of about 0.4 ppm(v) to 1.65 ppm(v), in the same period.

The world's surface temperature, which was about 4 C colder than today during the last ice age, 80 000 to 10 000 years ago, has risen about 0.5 C since the beginning of this century, most of the increase occurring in recent years. The sea level, measured in various spots all over the world, seems to have risen by 10 to 20 cm during the last century. It has varied between 100 m below and 300 m above the present level during geological times.

The problem is of course very complex, through the influence of many other factors, such as volcanic eruptions, wind and ocean circulation, sun spots, and even astronomical considerations. Thus, a consensus has not yet been reached. The evidence, however, seems to be pointing towards a relationship between CO_2 concentration, surface temperature and sea level. Current estimates point to a rise in temperature of about 4 C, and in sea level of about 0.4 m by the year 2100.

We are thus left with the following conundrum:

(a) The world's consumption of energy could be multiplied by 6 by the year 2100, calling for some 60 TW of additional capacity.
(b) The consumption of carbon-based fuels should not increase, and should in fact be curtailed.

Where will the additional energy come from?

2. Alternatives to Carbon Emitting Fuels

2.1 Solar energy

The radiant power coming from the sun to Earth is about 100 PW, or about 7000 times our present consumption. How can a significant fraction of this be captured? Possible mechanisms include the following:

- *Biomass*: the total photosynthetic storage by plants and trees represents about 150 TW; about one half of this is used to support plants' metabolism, leaving some 75 TW in biomass. Of this amount, about 1 % (500 MTOE/y, or 660 GW) is used for wood burning and a little more (700 MTOE/y, or 930 GW) for other purposes. If the total energy needs per person had to be satisfied from this source, each man, woman or child would have to find and burn *4 tons of wood per year*. One needs to take into account that current forests should be maintained and not depleted, as is now the case in Brazil, Indonesia and elsewhere.
- *Hydroelectricity*: this is the second largest sun-derived energy source, supplying about 3 % of the world's total needs. It has the advantage of causing hardly any pollution, and the disadvantage of covering large areas of land with water - typically 1000 km^2 per GW for an average dam. The total estimated exploitable potential has been set at 15 PWh/y (1.7 TW). This would clearly not be sufficient.
- *Solar water heaters*: there are small-scale applications, but no large project is considered feasible.
- *Photo-electricity*: this is a very promising avenue, depending on technological improvements. Currently, available cells have a low efficiency (15 %, maybe 19 % in the near future) and a high cost. A 1 GW plant would cover about 10 km^2, and would only

produce electricity (in the form of low-voltage direct current) in daytime, and, obviously, only when the weather is favourable.

- *Hydrogen (as a carrier)*: the scheme would be to produce electricity with photo-voltaic cells in sunny regions and use it to produce electrolytic hydrogen, which could then be piped to the consuming areas and burned to provide power (or maybe used in fuel cells). Technological developments are needed and the capital expenditure will be heavy.
- *Wind power*: windmills have been used for centuries, but have now been replaced with cheaper fossil fuel plants. Practical world capacity has been estimated at 2 TW, which, again, is only a fraction of the total need. Siting could pose problems, as well as the erratic behaviour of the wind.
- *Power from the sea*: three avenues have been explored: (a) the difference in temperature between deep and surface waters, (b) tides, as explored in the so-called Rance plant, in France, producing some 540 GWh/y (60 GW), and (c) waves. The potential of power from the sea appears quite limited.

2.2 Geothermal power

In parts of the world where the geothermal gradient is rather high, it may be economically worthwhile to pump water down to produce hot water or steam. A total capacity of some 6 TW(e) in electrical generation and 20 TW(e) in space heating and hot water supply has been installed since the 1950s.

2.3 Fission

The conversion of mass to energy in a nuclear reactor would seem to offer at least a partial solution to the CO_2 greenhouse problem. The first fission-generated electricity came on line in 1955. In 1995, nuclear reactors produced some 2.4 PWh(e) (8.74 EJ) of electricity, 18 % of the world's electricity and 7.3 % of the world's energy. Hereby, about one half to one million tons of CO_2 was avoided from being sent into the atmosphere, not to mention many other pollutants. Today, 43 new reactors are under construction, world-wide. More are being planned, but progress is becoming more and more difficult, through public and political worries, largely based on the three problems mentioned above, related to reactor safety, proliferation risks and waste disposal. These points are considered in detail in other parts of this book. At this stage, it suffices to state the following.

The Three Mile Island accident does not seem to have been responsible for any deaths. The casualty count for Chernobyl is less than the number of people killed per year in automobile accidents in the former Soviet Union - much less than the same figure for the United States. The number of deaths caused by the use of fossil fuels (from mining to pollution) has not been calculated. The risk of nuclear proliferation is limited. It has been argued that a small rogue state would probably find it cheaper and easier to develop and conceal the production of chemical or biological weapons, than to consider nuclear weapons. No final solution has been adopted for the disposal of radioactive wastes, but several are being studied. The time available to solve the nuclear waste problem is *not* essential: we can wait until all solutions have been tried and the best one has been determined.

2.4 Fusion

Fusion is the great hope for the future, since it could provide large reserves, little radioactivity, and no gases or ashes. The technology, however, proves to be rather intractable: a temperature of the order of 200 10^6 C must be maintained for some tens of seconds in a substantial plasma volume, at a pressure of a few atmospheres. Several ways are being explored, such as in the experiments Tokamak, Stellarator, and the "International Thermonuclear Experimental Reactor" (ITER), but cuts have been made in funding, and no commercial solution is expected to be developed for years - maybe not even before the end of the 21st century.

2.5 Costs

A comparison of costs between these different energy sources is fraught with difficulty, specially if "total" costs must be reckoned, i.e. including present and future damage to the environment (including the greenhouse effect), health hazards to the population at large, development costs, etc. Moreover, improvements in efficiency and safety will be capital intensive, baring the risk that the plants needed to provide the six-fold increase in energy production, by the end of the next century, will *not* be provided with all the necessary equipment.

2.6 The optimum Earth temperature

The question may be asked what the optimum Earth temperature is. Global warming might increase food yields at high latitudes. The rise in sea level due to ice melting, however, would raise many problems, such as flooding of lowlands in Bangladesh, Egypt and the Mississippi delta (where half a million people would have to be displaced), and the necessity to raise dikes all over the world. On the other hand, global cooling would lower the sea level and increase the land surface, at the cost of a decreased agricultural productivity at high latitudes. There is also a real, if quantitatively unknown, risk that an apparently quite moderate change in earth temperature might trigger a major climatic change: accelerated melting of Arctic ice might, for example, disrupt the North Atlantic Drift and plunge large parts of Northern Europe into conditions now current in Alaska and Northern Canada. To avoid armed conflicts, it would probably be better if the land temperature remained as it has been for the last ten thousand years.

Some Conclusions

The world population will continue to increase, but at a slower rate as peoples' living standards improve. The world population will flatten out around the year 2100 at probably almost twice the present level. Developed countries will increase their Gross National Product (GNP), but might keep their total energy consumption around the present level through economies, and increases in efficiency, in most countries. To raise living standards, decrease poverty and fight ill-health, the energy consumption of developing countries will have to increase, raising the total energy use of the world to about six times the present level by the year 2100.

In the beginning, developing countries will be mainly using fossil fuels to provide this additional energy. Since they are caught in the "poverty trap", this use will be inefficient and polluting. As time goes by, efficiency and economy will improve and the relative amount of pollution will be reduced, but this will be offset by the increase in total energy used.

The pessimistic, but probable, outcome, is that greenhouse gas emissions will rise and that the World's health will be impaired. Through the greenhouse gas effect, caused by CO_2, CH_4 and other gas emissions, the world temperature will increase. The rise in temperature will cause a modest (about 0.5 m) rise in sea level and bring about complex climate changes, disturbing established society patterns.

Natural gas, being cheaper and appreciably cleaner than coal, will gradually become the dominant fuel source. The share of nuclear fission reactors will continue to increase. Opposition might abate when pollution and dangers to health from the use of fossil fuels are better understood by the public and politicians. This will not be appreciated by everybody, however: e.g. miners who have spent their lives digging coal underground and who will have to be told that their product is harmful.

Nevertheless, fossil fuels will go on dominating the market for much of the 21st century, although depletion of oil and gas reserves may slow this down, at the end of this period. Coal will continue to be used until the 22nd century, when depletion might set in, but might be replaced by oil shale or tar sands, unless fission becomes widely acceptable or fusion becomes available.

No power supply based on solar energy looks as if it will be sufficiently abundant and cheap to be a major replacement for coal. Non-polluting hydroelectric power, even if fully exploited, will never cover more than 5 % of the world's total energy needs. Photo-electricity calls for expanded research.

Because of the dearth of other abundant and relatively low-cost energy sources, it seems that the conversion of mass to energy (fission and fusion) is the only alternative. Fission reactor technology is now well-developed, and modern reactors, with their relative reliability and good safety and health records, might substantially replace fossil fuel plants; it was poor design and irresponsible management that led to the Chernobyl disaster. The risk of nuclear proliferation, leading to the possibility of war, must be defused.

Fusion largely avoids these dangers, but research is still in a very early stage. It has not even been demonstrated that it could provide a practical energy source. Even if successful, it might not be available before the year 2100 - too late for greenhouse gas reduction.

The ecological movement has played an important role, particularly in insisting that *all* factors, including general health, be considered in decision-making. The movement needs, however, to define its goals more coherently and avoid internal dissension.

There is no easy solution. The rise in energy consumption by a factor of six (or maybe five or seven) *will* occur, assuming that a major war using weapons of mass destruction is avoided. Increase in pollution, higher temperatures, rise in sea level *are* to be expected. The seriousness of energy problems *must* therefore be recognised.

Energy savings, through better design and improved use, are a necessity. Public awareness is a factor, but political action, such as energy tax, suppression of subsidies to coal and oil production, and financing of research, is called for. Help should be given to developing countries, including Russia and China, to achieve - quickly - high efficiency, safety in operation and environmental friendliness in energy production. Hydroelectric production and solar power should be encouraged and developed. Ecologists should cease their internal discord and help in developing priorities and broad objectives. Nuclear energy should progress, with due regard to reactor safety, disposal of waste radioactive products, including plutonium, and avoidance of proliferation. Fusion research funding should be increased and different approaches tested.

In this chapter, I have tried to be pragmatic and use only assumptions derived from past experience, avoiding unreasonable "back-tracking" from an assumed sustainable future situation. The world population *might* stabilise around the year 2020, the energy consumption per capita *might* be reduced to 1.5 kW or be held down to 3 kW, there *might* occur a disastrous war or plague, the West *might* accept a substantial reduction in lifestyle, the South *might* renounce the improvement and progress related to increased energy use. These are, however, neither valid nor credible bases on which to build the future. In this matter, there are many technical considerations, but the answer is probably political. As the Rio and Kyoto conferences have demonstrated, very few world leaders have other than short-term - i.e. next election - goals. Very few of them are real statesmen willing to look ahead to long term energy and population problems.[45]

[45] The author acknowledges helpful comments and discussions with J. Avery, H. Drevermann, M. Graetzel, G. Harigel, M. Hine, J. Holdren, S. Kapitza, R.S. Pease, A.H. Rosenfeld, E. Velikhov and F. Waelbroeck. They are, however, not responsible for any of the opinions expressed. Thanks are also due to G.R. Grob and H. Sharan of the World Circle of Consensus for inviting me to talk at the 1991 World Clean Energy Conference. This chapter has no connection with CERN.

Chapter 4 Energy Efficiency is the Key

by
Benjamin Dessus*

Introduction

When we attempt to anticipate the energy problems facing mankind in the twenty-first century, beyond the years 2010 or 2020, we can examine things from different points of view. First, from the classical point of view, that of energy availability, where one tries to extrapolate, starting from today, the evolution of the markets of the main energy resources as a function of technical progress, evaluation of reserves, geo-strategic considerations, etc. This is the art of the forecaster: a perilous art, and the more so as one projects ahead in time. Beyond a few years, reference to the past no longer suffices and a simple continuation of observed tendencies runs the risk of being contradicted by facts.

But we can also start from the other end, from the development needs of societies, which leads to the long term world energy demand, say in the year 2050 or 2100. From there, one can build up backwards towards the present time. This leads us to a prospective approach, in which we try to build contrasted, but coherent pictures of distant horizons, and we consider first, not energy, but the set of challenges and risks which almost inevitably will confront humanity in the forthcoming hundred years. Such are the future challenges which energy systems will have to accept in order to make a positive contribution to development. How can we face the challenge of an equitable development for the 10 billion people which will populate the planet in the year 2100, of supplying them with the necessary energy services, without however multiplying global risks for humanity?

Of course, this global challenge involves a whole set of issues, ranging from sufficiency of food supplies to housing and health; ranging from the respect of human rights to democracy, from education and culture to the peaceful resolution of conflicts, from the protection of childhood to the maintenance of the large scale ecological equilibrium of the planet. The question raised concerns the contribution of the most judicious energy strategy to the positive solution of a wide range of challenges. Of course, energy does play a major role in all the aforementioned challenges. It is central to some of them and it contributes almost always, at least partially, to the solution of most of the others. Energy systems are one of the ways to face these challenges but they do not constitute a goal as such.

But more generally, the importance of the services provided by energy in most of human activities gives a special importance to economic efficiency when choices are

---

* Benjamin Dessus, engineer and economist, is Director of the Interdisciplinary Research Programme on Technologies for Eco-development "ECODEV" at the French CNRS (*Centre National de la Recherche Scientifique*). He is President of the *Conseil scientifique et technique pour l'environnement mondial*, as well as President of the Association "Global Chance".

made. Energy systems have to be conceived as a means for providing a certain number of energy-requiring services, and not solely as efficient means for producing and distributing energy. Rigorous and equilibrated choices are essential because of the importance of the invested capital and of its long period of immobilisation. Amory Lowins from the Rocky Mountain Institute reminds us: “Wasting energy is not just a minor error in the context of development. In fact, inefficient motors or lamps steal money which could have been used to provide drinkable water, vaccination of children or equality of women”.

If we aim at providing energy services to ensure the development of human societies under the best possible conditions in the next century, it is of particular importance to evaluate the extent of the associated constraints and risks. The analysis of the constraints on the development of energy systems raises a methodological problem. First, one must avoid both the temptation of viewing long term development from a solely energetic point of view (at the risk of omitting major considerations about the evolution of societies and of the life conditions on the planet), and also that of considering the energy constraints as purely exogenous. The second difficulty is linked to the concept of “global risk”, which lies at the heart of the notion of “global change” and its relation to time. Energy is often involved even if it does not bear the full responsibility.

This “global” notion, which spread towards the end of the 1980s in connection with major environmental problems (the heating up of the atmosphere due to the greenhouse effect, the undermining of bio-diversity, desertification, etc.), spans both a spatial dimension (the problem is global insofar as it affects directly the full set of human societies, and therefore the planet) and a temporal dimension insofar as it affects the concern of humanity for its future generations. In fact, because of this concern for the protection rights and resources of future generations, the global notion moves from its spatial meaning towards that of a “human heritage”. This reference to “global risk” leads to the existence of a “concern involving all humanity.”

How then is one to face simultaneously the four risks which are linked to energy and which present a global character to humanity:

- the risk of rarefaction and exhausting of fossil fuels (coal, oil, natural gas),
- the risk of warming up the climate due to intensive use of fossil fuels, the reality of which was confirmed at the Kyoto conference on Climate Change in 1997,
- the nuclear risks, both civil and military (accidents, transport and stockpiling wastes, the risk of proliferation) of which we have been reminded by recent events[46],
- finally, competition for the use of land would run the risk of a too intensive use of land, otherwise suitable for cultivation, for the purpose of energy production.

---

[46] In particular, the stockpiling of very long lived radioactive wastes presents a global character in that it concerns a large number of future generations (several thousand, if not tens of thousands of generations). The consideration of such time scales is quite exceptional in human activities and it probably explains the special sensibility of public opinion for this issue. On the other hand, the risk of leakage to the biosphere is generally considered small and limited to a small surface area.

We immediately face a first difficulty: indeed the consequences of such risks are not really felt by the population today, nor are they accurately measured by scientists. On the other hand they are not independent; thus, for example, in order to fight against global warming, we could resort to a reduction of the use of fossil fuels. In doing so, we push forward the date at which fossil fuel reserves will be depleted, while at the same time we mitigate the emission of gases which enhance the green-house effect, responsible for global warming. But if we resort to an increase in the use of nuclear power or to renewable energy sources in order to face these needs, we increase by the same amount the risks associated with nuclear power and with competition for the use of land. It is therefore vital to keep in mind this interweaving of medium and long term risks, if one is to define an energy strategy aimed at a lasting development.

Finally, their consequences may take the form of radical disruptions: even if the depletion of fossil resources is a continuous phenomenon, the main difficulties will arise less from the long term predictable exhaustion than from tensions which will occur in the oil market and even from regional conflicts. The same holds for the change of climate.

1. Development and energy demand

Before delving into a prospectus of energy resources, of their extent, of the technical and environmental issues which they raise and which constitute the main concern of energy specialists[47], it is vitally important to try to reach a better understanding of the nature of the links between development, or more restrictively economic growth, and energy demand. This is of course a strategic element of the prospective study.

For a number of years, the relation between the consumption of energy and economic activity has been formulated by a model which assumed an 'elasticity' equal to one: the ratio of energy consumption to GDP (Gross Domestic Product) was assumed to be constant.[48]

During the 15 years which followed the first oil crisis, the industrial countries experienced a de-coupling of economic growth and energy consumption (economic growth was maintained alongside constant energy consumption per capita). This reduction of the energy intensity (which however practically ceased in the northern countries following the 1986 oil counter-crisis), which was interpreted for some time as a simple adaptation to new economic variables, appears rather to be rooted in two structural phenomena. The retrospective analysis of the development of industrialised countries, and the study of the more recent evolution of developing countries, have led to progressive abandonment of the first model in favour of representations with variable elasticity, which take into account the fact that the consumption of energy depends on the level of development and on the state of technology (see figure 1).

The historical analysis of the development of industrialised countries has shown that all countries have witnessed an increase of their energy intensity during the first stages of industrialisation, followed by a stabilisation and an eventual decrease (see again

---

[47] In the original French version of this text the word "*énergéticien*" was used.

[48] The ratio of energy consumption to GDP is the energy intensity.

figure 1). But this evolution covers contrasted evolutionary sectors. For example, a European consumes today about 3.5 TOE (ton of oil equivalent) per year, against 2 TOE in 1960. It is interesting to examine the evolution of the consumption structure. The needs in 1960 were dominated (70 % of the total) by consumption related to the servicing of houses (the development of thermal comfort) and to industrial production. The consumption in these sectors has decreased in absolute value since 1973 and today it represents only half of the energy consumption. The present economic growth involves only a very small growth in these sectors. An important fraction of new requirements are replacements of old demands: increased living space is more than compensated for by improvement of energy performance.

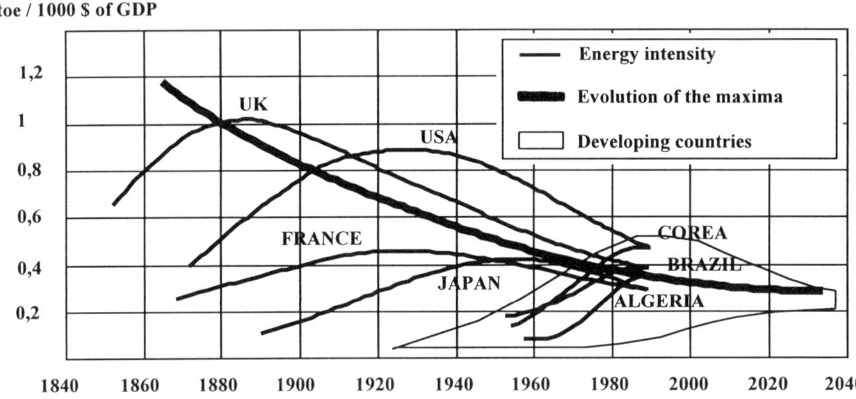

Fig.1. Evolution of the energy intensity of industrialised countries during the past one hundred years.[49] *Source: B. Dessus, Energie: un défi planétaire, Belin, Paris, 1996.*

On the other hand, other groups of users have witnessed a sustained development during the studied period (manufacturing industry, food, lighting, white electric household goods) thereby causing an important increase of energy consumption in these sectors. But the latter are less intense energy consumers than the former: they represent only about 20 % of the energy consumption of a European.

Finally, only three categories of uses have shown and still undergo a sustained increase of demand: specific electric consumption (the "brown" domestic sector, but

[49] The thick solid curves represent the energy intensity evolution of 7 (largely) developed countries: Algeria, Brazil, France, Japan, South Korea, the UK and the US. The single "fat" solid slope shows the evolution of the maxima of these curves. The thin line represents the envelope of the energy intensity evolution of various developing countries. The figure shows that recently industrialised countries develop with a lower energy intensity than their predecessors, since they benefit from the technological progress acquired. The dollar value used is that of 1980.

mainly the equipment of service industries), transportation of people and of goods. Among these, the transportation sector weighs heavily on the energy balance (consumption has increased by a factor of two since 1960).

But the evolution of energy demand also bears the mark of infrastructures. While industrial production in France accounted for 40 % of the energy consumption in the 1960s, it accounted barely for 30 % in the 1990s. The remaining 70 % depends strongly on heavy infrastructures such as urbanism, real-estate and transportation networks, most of which have a lifetime which exceeds a half a century. It is thus by taking a careful account of these considerations, and in particular of the rhythm of technical progress, of infrastructures and ways of life and organisation, that one must construct a prospectus of energy demand.

2. Scenarios for the Future

For a long time, energy specialists, who are well aware of the large inertia of energetic systems, have adopted the habit of projecting into the far future, if only to anticipate investment needs. In this tradition, in order to tame the future, numerous teams, related to universities or close to energy producers, have recently drawn contrasted prospective images of energy in the world in the years 2020, 2050, and even 2100. When we observe the scenarios which they offer, we quickly notice that they can be grouped into two categories:

The first scenarios propose a vision of the future constructed on a productivity model of "development through abundant energy", with contrasted options regarding the contribution of different primary sources. Such scenarios admit risks as being inescapable ("one cannot make an omelette with breaking eggs"), they pile them up to a high level and they differentiate themselves by the increase or decrease of one risk with respect to the other three. In order to avoid being too hot (the green-house effect), claims one, let us accept nuclear risk. No, replies another, I prefer being too hot rather than leaving the problem of nuclear wastes to my grand-children. In fact these scenarios quickly run into numerous contradictions: how can one guarantee easy access to fossil fuels if their consumption is to increase very fast, in a context of unequal economic growth scarcely favourable, a priori, to the development of world trade? How is one to face the environmental tensions which can arise from a high dependence on fossil fuels, an important production of nuclear wastes and, for some, from a massive reliance on renewable resources?

The second group of scenarios propose a "development by energy efficiency", which attempts to re-equilibrate energy policy by giving a strong priority to the control and evolution of energy demand. Starting from a detailed analysis of final energy needs of development, they express the will to simultaneously push forwards in time the main risks and to favour the development of southern countries while anticipating a reduced use of capital for the development of energy systems. They assume a deep cultural revolution because they aim at a strong de-coupling of economic growth and energy consumption. In particular, they imply a broadening of the range of intervention of energy policy toward the whole sector of activities which creates demand (transportation, construction, urbanism, equipment goods), because, as we have seen, the determining factors of energy demand are often closely linked to the large infrastructures of urbanism, transportation, and various networks of distribution of fluids (energy, water, etc.).

In the long run, the studied scenarios distinguish themselves more by the volume of the energy demand than by the assumed mix of energy sources. The group of "energy abundance scenarios" assume, already in the year 2050, an energy consumption 3 times higher than the present (22 to 25 billion TOE per year, whereas the "efficiency" scenarios can manage with 12 to 15 TOE per year. Fifty years later, in 2100, the divergence between the scenarios exceeds a factor of 5 (see figure 2).

From the point of view quoted above, only the "efficiency" scenarios appear to be capable of avoiding major disruptions for humanity, insofar as they postpone the dates of most risks by a good thirty years (see figure 3). Scenarios which assume a strong decrease of global world energy intensity yield a most significant solution regarding global risks, because they allow us to minimise strongly the set of risks at the horizon of 2050. This is shown in figure 3, which compares seven scenarios: four "abundance" scenarios (A1, A2, A3 and B) of the International Institute for Applied Systems Analysis (IIASA), conducted by the World Energy Council (WEC)[50], and three "efficiency" scenarios, two from IIASA (C1 and C2) and one, called Noé[51], from the French *Centre National de la Recherche Scientifique* (CNRS).

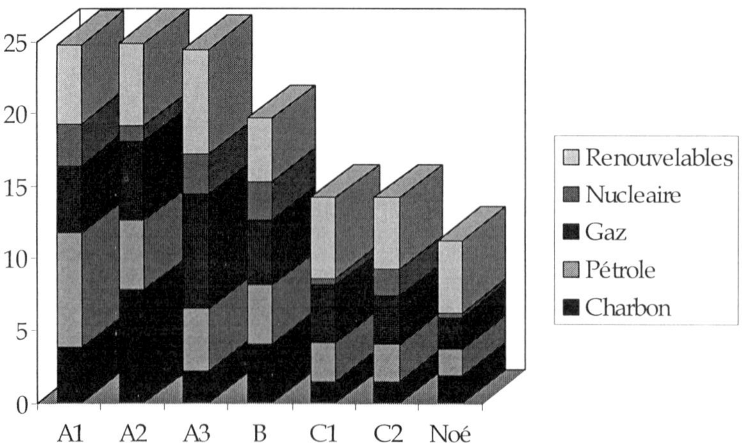

Fig. 2. Two contrasted scenarios: Abundance (A1, A2, A3 and B) versus efficient energy (C1, C2 and Noé) for development.[52]

[50] *Global Energy Perspectives to 2050 and Beyond*, World Energy Council, IIASA Report, 1995.

[51] "Jérémie et Noé, deux scénarios énergétiques mondiaux à long terme", Benjamin Dessus and François Pharabod, *revue de l'énergie*, n°421, 1990.

[52] The original histogram of this figure was depicted in colours. The shadings used here, for each of the 7 scenarios, going from top to bottom, correspond to renewables (*renouvelables*), nuclear (*nucléaire*), gas (*gaz*), petrol (*pétrole*) and coal (*charbon*).

In addition, they compare favourably with the abundance scenarios from the economic point of view. This economic advantage can be largely explained by the fact that production and energy distribution costs often exceed the cost of saving energy.

The lesson to be drawn from these exercises is clear: by basing economic growth on a strategy of energy efficiency, humanity can avoid, without causing ruin, the unacceptable wager of replacing one risk for another, or that, equally dangerous, of waiting for a technological miracle which would cheaply spare the planet and its inhabitants from the above-mentioned risks. Answers to the energy challenge in the mean and long term, based solely on technical progress applied to energy-supply technologies, are unable to meet the various challenges of development, of resources and of the protection of the environment. It also does not appear acceptable to rely solely on the appearance of technological breakthroughs in order to solve these problems.

	A1	**A2**	**A3**	**B**	**C1**	**C2**	**Noé**
Carbon Accumulation (10^9 tons)	545	610	470	488	356	350	320
CO_2 Concentration Increasing	40 %	50 %	33 %	35 %	23 %	23 %	20 %
Oil Resource Exhausting	104 %	85 %	81 %	75 %	60 %	60 %	50 %
Nuclear Accident Risks 1990-2050 (100 for A3)	95	50	100	95	45	72	30
Nuclear Waste Stock (100 for 1990)	1700	900	1800	1700	800	1300	550
Land Competition (10^6ha occupied by renewables)	800	800	1300	650	800	700	800

Fig. 3. The impact of various risks for different scenarios in the year 2050.

On the other hand, if the world can base its economic development on energy efficiency, it can avoid the dilemma, unacceptable today, of trading one risk for another and it can contribute to lowering the tensions arising between and within societies. There is a double interest in this approach: on one hand, progress in our knowledge will help us better to evaluate the extent of the risks and of their relative importance; on the other hand, technical and industrial progress will lead to new, and sometimes unexpected, technological opportunities for supply and demand at the level of research and development (R&D).

The protection of the environment does not therefore translate into economic recession; there is no systematic contradiction between "development" and "sustainability", but, instead, a profound synergy. Strategies with low energetic profile are win-win-strategies, both on economic and environmental grounds.

It is therefore an optimistic message which emanates from these exercises. The real question facing us today is more which strategy to adopt, in order to set course in the direction suggested by the sober development scenarios, rather than to dissipate most of our strength in a spiral which consists in developing, always faster and without limit, both energy systems and technologies that are capable of limiting their potential for damage.

3. How to do it?

Energy producers are, of course, not the only ones concerned by this question. Indeed, we know that the energy consumption associated with the development of societies depends mainly on infrastructure decisions (transportation, urbanism, etc.) which are taken every day; they determine irreversibly the nature and the quantity of energy required to satisfy a given service for great periods of time, and they relativise technical progress all the more. When we consider for example, that today 70 % of energy consumption in France arises from dwellings, services and transportation, we realise the importance of the choice of infrastructures in the French energy balance.

Public, national and local authorities responsible for land management and democratic debate are at the very centre of the choice of long term policies, which the market is unable to take into account.

As for energy producers, they too have to undergo a cultural revolution, which has barely begun today, in order to make the transition from a pure energy supply logic to a logic aimed at satisfying the ultimate aims of the consumer at minimal energy cost.

Finally, for industrialists who manufacture materials, be it cars, boilers or refrigerators, the challenge is precisely to offer the population of our countries sober and clean materials which will permit a lasting de-coupling between the economic growth of our industrialised societies and the growth of energy demand and, thereby, of environmental problems.

Without such a threefold example being set by our rich societies - an effort arising from public authorities, energy companies, industrial producers of goods and services - how can one expect that developing continents such as Asia, Latin America or Africa will, in the twenty first century, adopt modes of development which will not seriously compromise the future of mankind on the planet?

Thus, the challenge we are facing is not technical: it is cultural, it is political. If ecological horror can be avoided, it is up to us, citizens of the world, to react, and to develop tools of a planetary solidarity with which to build an equilibrated and sustainable development of human societies as a whole.

Reference

B. Dessus, *Energie: un défi planétaire*, "Collection Débats", Editions Belin, Paris, 1996.

Chapter 5 Problems and Prospects for Nuclear Power in India

by
P.K. Iyengar*

Introduction[53]

India's nuclear programme was based on the perceptions of two persons: Homi Bhabha, who was an outstanding scientist and contributed to the progress of nuclear science from the early 1930s to the 1960s, and Prime Minister Pandit Jawaharlal Nehru, who strongly believed that the nation could progress only through the introduction of science and technology into its fabric.

Homi Bhabha was a basic scientist and he knew that in order to establish a new technology one needs to start with knowledge generation, technology acquisition and manpower training, as three essential components, even though resources may also play an important part. With the strong belief, arising out of the Indian culture, in the basic capability of India in science and technology, he developed a programme which would have been unimaginable for a developing country to embark upon.

1. Milestones

The appendix to this chapter lists the milestones in the development of nuclear science and technology in India. What one notices in this is the concurrent development of important facilities, technologies and innovations consistent with its own objectives. While in the early period it was essential to develop manpower in basic research, this called for the establishment, over many years, of expertise in many disciplines, as well as for introduction of new blood into the system. A training school for graduates in science and engineering was started in 1957. About 200 high quality candidates were selected and trained by scientists and engineers in the Bhabha Atomic Research Centre (BARC)[54] for one year, and then inducted as research staff. Over the last 40 years this has helped to introduce more than 6 000 researchers of high calibre into the system.

---

* P.K. Iyengar is a former Chairman of the Indian Atomic Energy Commission.

[53] This is a slightly shortened version of the original paper by Dr. Iyengar.

[54] BARC: Bhabha Atomic Research Centre.

The research reactors, programmes for materials development, instrumentation using the most modern electronics, chemical engineering processes connected with fuel fabrication, enrichment and fuel cycle, were all achieved as a part of the in-house programme rather than by the mere acquisition of technology from outside. This can be seen in the milestone chart from 1955 to 1965. The strength of the Indian programme is entirely different in conception and mode of operation to that in many other developing countries. It is no wonder therefore, that India had the capacity to detonate a nuclear device in the early 1970s, and has now graduated to making a thermonuclear device. It was able to make enriched uranium by the centrifuge method in the early 1980s, experiment with new fuel (plutonium carbide) in fast reactor systems, and make ^{233}U easily available for the future development of atomic energy. In the milestone chart one can see a continuous evolution of these capabilities irrespective of the fact that external assistance, which triggered the use of atomic energy for peaceful purposes, ended in the 1970s after India's first test of a nuclear explosive.

2. Nuclear Power Programme

Fission energy can be released from ^{235}U, ^{239}Pu or ^{233}U in a variety of ways depending upon available resources and technologies, and on the economy. Some industrially advanced countries, (France, the UK, Canada), believed in the use of natural uranium/heavy water as the best fuel/moderator combination for nuclear reactors; this process was efficient enough to produce nuclear power economically in the late 1950s and 1960s. The development of the low enrichment uranium/light-water reactor, derived from military applications (submarines) allowed the US to offer less expensive technology transfer to the nuclear industry of the advanced nations, together with an attractive price for the fuel (the technology developments having been paid for by the defence budget). Canada retained the pressurised heavy-water reactor as a reliable, economic, and scientifically logical method of getting a maximum of energy from a given quantity of uranium. India followed the same principle, maybe due to the friendship and mutual regard that developed between Dr. W.B. Lewis of Canada and Homi Bhabha of India when they were together at school in England.

The three-stage programme for the development of nuclear power in India was formulated by Homi Bhabha in the late 1950s. As shown in table 1, in the first phase this consisted of building a large number of pressurised heavy-water reactors using natural uranium-oxide as fuel, and heavy water as coolant and moderator, similar to the Canadian so-called CANDU reactors. It was planned to establish nuclear power stations up to a capacity of 10 000 MW using local sources of uranium. Taking advantage of the plutonium that these reactors would produce, with a conversion ratio of about 0.8, India would then build fast breeder reactors, producing $^{234}{}_{94}Pu$ (or $^{233}{}_{92}U$, if thorium is used in the blanket). With the unlimited resources of thorium available in India, it would then allow a multiple choice of thermal and fast reactors which would meet India's requirements for several centuries. In spite of several obstacles, this basic programme is being followed, and excepting for changes arising out of resource limitations and technological innovations, the basic approach remains the same.

Stage-I	Construction of natural uranium (heavy water moderated) Pressurised Heavy Water Reactors (PHWRs). Spent fuel from these reactors is reprocessed to obtain plutonium.
Stage-II	Construction of Fast Breeder Reactors (FBRs) fuelled by plutonium produced in Stage-I. These reactors would also breed ^{233}U from thorium. It is now planned to develop an Advanced Heavy Water thermal Reactor (AHWR), as an extension of Stage-I PHWR programme. The AHWR, using a ^{239}Pu enriched uranium fuel in the driver (booster) zone and ^{233}U enriched thorium fuel in the driven (blanket) zone, would generate a large part of its energy output from thorium through fission of *in situ* bred ^{233}U. Thus accelerating utilisation of thorium.
Stage-III	Power reactors both thermal and fast using ^{233}U/thorium as fuel.

Tab. 1. The three stages of the nuclear power programme of India.

3. Lessons Learned

Several lessons have been learnt from the Indian effort during the first three decades:

(a) India was successful in developing manpower rather quickly and in sufficient numbers in innovating areas, like, for instance, the synthesis of plutonium carbide fuel for the Fast Breeder Test Reactor (FBTR) programme. However, a different approach was required when it came to "large-scale technology adaptations", like high-energy particle accelerators, large-scale isotope separation and reprocessing. The absence of a local industrial engineering tradition slowed down many of these developments, even though they ultimately got established. For example, it took a couple of decades and the building of a chain of reprocessing laboratories and plants to establish the intricate technology involved in the large industrial-scale production of heavy-water. It took equally long to develop the industrial production of nuclear power station components. The special needs of the nuclear industry called for the introduction of new concepts, like quality control and non-destructive testing of basic materials. BARC had to carry out pilot plant studies and build small production facilities before scaling up to large production establishments. Examples are:

- the Nuclear Fuel Complex, which manufactures zirconium alloys, nuclear components like pressure tubes, and all kinds of nuclear fuel for power reactors;

- the Electronics Corporation of India Ltd., which manufactures electronic items from control systems to computers;
- the Uranium Corporation of India, which mines and processes ore to yellow cake;
- heavy water production plants, which now deliver more than 600 tonnes of heavy water per year;
- an industrial-scale centrifuge separation plant for enriching uranium.
- the Board for Research in Isotope Technology (BRIT), which supplies all sorts of pharmaceuticals and isotope sources on an industrial scale for use in health, agriculture and industry.

(b) The development of exotic materials and new systems (zirconium alloys, improved heat exchangers, steam generators or turbine generators) made tremendous demands on an unwilling industry used to colonialism. Usual technology transfers were no more valid; suppliers' cartels, arising from the 30-year old Non-Proliferation Treaty (NPT) regime, created problems. How India survived these sanctions, what relevance these sanctions had to concepts in nuclear proliferation, and how it uniformly affected all developing countries, whether or not they were signatories to the NPT, are now facts of history.

(c) For a start, India bought two boiling light-water reactors from the US in 1964. These were of the first generation nuclear power stations offered commercially in the early 1960s. They were installed in the fairly short time of six years, but the co-operation with the US for the supply of spares, components and fuel came to en end in 1974. India had to develop its own substitutes to continue operating these reactors safely, economically and for a longer period of time than the same reactors operating elsewhere. Even after twenty-eight years, these two reactors still produce electric power at a competitive price, and the operating company makes profit out of them. Even though similar reactors have been shut down in other parts of the world, India will continue to operate them safely for at least the next decade. This teaches us how a developing country can absorb and develop advanced technologies in spite of a lack of co-operation, and even an active dissuasion, from developed countries.

(d) It was almost at the same time, in 1964, that it was decided to co-operate with Canada and build 200 MW PHWR reactors, concurrently with their development at Douglas Point. The first of these was built at Kota in Rajasthan. The Douglas Point reactors of that type have since been closed down, but the Rajasthan reactors are continuing to operate after reconditioning core materials like the pressure tubes. The story of the problems that India had to solve in these two reactors, and how they did it, is itself a lesson for many developing countries. Co-operation with Canada ended in 1974; since then, India has built and commissioned six PHWR units of 220 MW each, with gradual development of indigenous technology, and even changes in its main features.

India has since started building an upgraded version of this reactor: two 500 MW PHWR reactors at Tarapur, in addition to four 220 MW units, which will be commissioned in the next couple of years. An advanced heavy-water reactor system is under design, which will incorporate new safety features like a pump-free natural circulation cooling system, and the use of a mixed thorium/uranium core for better conversion ratio and energy production per kilogram of uranium. A list of power stations under operation and construction is given in Table 2.

Plants under Operation	**Initial Criticality**
2 x 160 MWe BWR [a)] at Tarapur	1969
1 x 100 MWe PHWR at Rawatbhata	1972
1 x 200 MWe PHWR at Rawatbhata	1980
2 x 170 MWe PHWR at Kalpakkam	1983, 1985
2 x 220 MWe PHWR at Narora	1989, 1991
2 x 220 MWe PHWR at Kakrapar	1992, 1995
Plants under Construction	
2 x 220 MWe PHWR at Kaiga	
2 x 220 MWe PHWR at Rawatbhata	
Future Plans	
TAPP-3 & 4, 2 x 500 MWe: Construction to start soon.	
2 X 1000 MWe VVER at Kundankulam: Detailed project report under preparation in co-operation with Russia.	
4 X 220 MWe PHWR units at Kaiga and 4 X 500 MWe units at Rawatbhata: Projected for the future.	

BWR: Boiling Water Reactor; PHWR: Pressurised Heavy Water Reactors; VVER: Russian abbreviation for "pressurised normal water reactor".

Tab. 2. Indian Power Reactors under Operation and Construction.

4. Environmental Aspects and Technology Development

A definite correlation between social development and energy consumption per capita has been established by many people. The largest and most populous countries in the world, China, with about one-fifth, and India, with about one-seventh of the human population, consume around 500 kWh/y and 350 kWh/y of electricity per capita respectively (57 and 40 W/capita). Sustainable development of the entire world depends on the availability of adequate energy to this huge population. If the additional requirement to reach consumption rates of 2000 kWh/y per capita (230 W/capita) is met purely through fossil fuels, the impact on the environment through the release of wastes, particularly CO_2 and other green-house gases, would be catastrophic. This was highlighted at the Rio conference and in the more recent Kyoto meeting.

However, when it comes to alternative methods for energy generation, there is a lack of emphasis on the ability of nuclear power to provide an economically viable and environmentally acceptable alternative. This is due to several reasons: (1) the too rapid introduction of nuclear power, leading to near term saturation in some of the advanced countries and the sending of wrong signals; (2) the impractical unit size of 1000 MWe on the average, which cannot be easily connected to existing power grids in developing countries; (3) the dependence on enriched uranium fuel for these reactors, the sources for which are restricted and rather constrained by international conventions like the NPT, which infringe on national sovereignty; (4) the effect of the Chernobyl accident, which has put in question the reliability and safety of nuclear power in the hands of people who are used to only medium level technology.

Many papers have been written on the desirability and growth scenarios of nuclear power in developing countries. They have often tended to be unrealistic and unhelpful in the promotion of this alternative for sustainable development.

5. Other Nuclear Energy Production Techniques

5.1 Subcritical systems

Since the discovery of neutron-induced fission, this process has remained the principal method for the production of nuclear energy. But nuclear physics does not demand that the system be critical or self-sustaining in order to produce power. "Sub-critical systems", free from explosion dangers, together with an external neutron source providing a neutron flux appropriate to the needed energy density in the fuel, could be a viable and safer alternative. This has been proposed several times in the past, as by Canada in the so-called "ING" project, and is now being pursued vigorously by Carlo Rubbia and Los Alamos (see Chapter 16, by van der Zwaan).

5.2 Fusion

Fusion of light elements is another alternative which may avoid the creation of high levels of radioactive waste and the consequent problems. Fusing hydrogen isotopes (deuterium ${}^{2}_{1}H$, tritium ${}^{3}_{1}H$) to produce instantaneously an enormous quantity energy (equivalent to a few megatons of chemical explosive[55]) was demonstrated way back in 1952 in the "Mike" explosion. The controlled release of energy on a semi-permanent basis is, however, another problem, which, in spite of the progress made since then, does not seem to provide a solution in the foreseeable future. Closeness to nuclear weapons technology has wrapped up this field in secrecy.

[55] 1 equivalent Mt of TNT is equivalent to 4.2 PJ.

5.3 Fast breeder reactors

A breeder reactor is one that is designed to use extra neutrons from the fission reaction of a fissile element ($^{235}_{92}U$, $^{244}_{94}Pu$, $^{233}_{92}U$) to convert a "fertile" nucleus ($^{238}_{92}U$, $^{232}_{90}Th$) into a new fissile nucleus. The aim is to produce more fissile material than is being consumed and to allow the use of the abundant fertile elements for energy production. The fast reactor is one such device which has almost reached commercial size (Super Phoenix). While not being beyond the means of the future, the technology of fast reactors is certainly complicated. In spite of the significant advantages (lower radiation levels, reduction in radioactive wastes, higher steam temperature and higher efficiency), the process – dealing as it does with plutonium and highly enriched uranium – is meeting with considerable opposition, partly related to proliferation fears.

India started a fast reactor programme in collaboration with France in the early 1970s. FBTR, a 40 MW test reactor, has been built and is operating at Kalpakkam. It uses plutonium carbide as fuel, and has reached a burn-up level exceeding 30 000 MWD/ton without any problems. A 500 MWe Prototype Fast Breeder Reactor has been designed and constructed. Testing of the required components is in progress. India has decided to move further in this field.

6. How Market Forces Can Impede the Development of New Technologies

There are many areas of industrial activity which are restrained at present because of market forces rather than for scientific and technical reasons. An already established large-scale industrial base can prevent the introduction of new technologies which would be useful to the development of countries with a large population. This might be the reason why light-water reactors have not penetrated the developing countries' markets. Additional safety devices to the existing 1000 MW nuclear power plants in the developed nations have increased the capital cost of nuclear power beyond that of any fossil fuel source. This has resulted in unhealthy competition. The anti-nuclear lobbies in the media, especially after the Chernobyl accident, have struck fear in the governments and the public at large, mostly through fear of the unknown, which cannot be erased easily. This is our predicament, and unless a well experienced scientific community speaks openly about the advantages that were foreseen by the pioneers, their dream that "there will be no need to meter electricity distribution" will never come true.

7. Influence of Capital

Many developing countries (South Korea, Taiwan) have quickly increased their electricity generation by installing light-water reactors purely on a commercial basis. Nuclear power stations have become commercial products, but there is no technology transfer for fuel fabrication, replacement parts or services. This involves the call on foreign capital and the outflow of foreign exchange. For India, it would be foolish to depend upon imported fuel. The electricity production of the country should grow by at least 15 % a year to raise the average production from 350 kWh/y per capita (40 W/capita) to something like 2 000 kWh/y per capita (230 W/capita) by about 2020. Should this quantum jump in electricity production be dependent upon the import of equipment and fuel, a major economic disaster would be the consequence. The recent developments in South East Asia clearly demonstrate the inadvisability of foreign capital investments disproportionate to a nation's saving capacity. The Indian nuclear power generation programme had slowed down in the last ten years. With a firm conviction in national capability, and a renewed faith in self-reliance, the present government has already decided to accelerate investments in the nuclear power sector.

7.1 India

India and China became independent almost simultaneously. India established its Atomic Energy Commission in 1948, and developed all the infrastructure for introducing nuclear power, according to the milestone chart in the appendix. During the early years there was no attempt to produce enriched uranium having a direct relationship to nuclear weapons. The reprocessing technology for producing plutonium was however a mere chemical process which could be developed and automated with adequate confidence. Plutonium was chosen in view of the future development of fast reactors and thorium utilisation.

India had the help of many Western powers, starting with France in the early 1950s and particularly Canada later. It developed the manpower and regulations for the control of radioactivity and safety systems, in accordance with Western technology. The organisations have more or less the same structure as in democratic countries. There is an independent Atomic Energy Regulatory Board (AERB) which ensures safety and reliability in nuclear operations, including the use of radiation in industry, agriculture and health. It has maintained a remarkable safety record so far.

7.2 China

China, being a member of the communist block, received considerable support, training and technology from the Soviet Union, which supplied a research reactor, an enrichment plant, and other equipment towards the development of nuclear technology. China decided to acquire nuclear weapons and detonated its first atmospheric explosion in 1964. It obviously needed further enhancement of its capabilities for enrichment and nuclear weapons technology. Until about 1980 no effort was made to develop nuclear power reactors. China ordered two commercial light water power reactors from France, which have now been set up at Daya Bay. Simultaneously there was a programme to build more light-water reactors. A 300 MWe reactor was designed and built with some of the basic components imported from abroad, which went into operation in 1992. China is building a similar type of reactor at Chasma in Pakistan. It now has contracts with Germany, as well as Russia, to build 1000 MWe light-water reactors. It is also installing heavy-water reactors from Canada. The object is therefore mainly to be proficient in a number of reactor systems which can be introduced in the right proportion in the future. It obviously has heavy water production facilities, material testing reactors, production facilities for tritium breeding from research reactors, etc. The absence of a large-scale industrial infrastructure has been the only reason for importing large size nuclear units from abroad. The most recent US-China co-operation agreements will ensure the introduction of light-water reactors of US origin. The contribution of electricity from nuclear reactors is still small, around 1.27 %, but they hope to increase it in the near future.

7.3 Indonesia

Indonesia started in a big way to build the infrastructure for research in nuclear science after the Atoms for Peace initiative. It created research centres in Bandung and Jakarta, with research reactors and a number of laboratories for isotope production and distribution. In the mid-1980s it expanded its programme by installing an imported 40 MW research reactor from Germany. The scale of operations was expanded and a consultant was appointed to advise on the introduction of nuclear power. The consultant's recommendations are under study by the government, but no decision to order nuclear power stations has yet been taken. The options seem to be in favour of the light-water reactor or the PHWR. A regulatory board for safety has already been set up, and depending upon the financial resources, the country may venture on a nuclear power programme. Some sites have been identified for the power stations. The availability of electricity per capita, at 350 kWh/y per capita (40 W/capita) is low, and is also unevenly distributed in the vast stretch of the country. Nuclear power stations of very large capacity cannot be introduced because of the limited grid capacity.

7.4 Pakistan

Pakistan entered the nuclear field by about 1964, by installing a research reactor and contracting with Canada to build a 125 MW PHWR at Karachi. That reactor was commissioned in 1972 and Pakistan has had the experience of operating a PHWR system since that time. However, indigenous capability has not grown, because of insufficient industrial development. More recently, in collaboration with China, it is building a

300 MW light-water reactor. In other techniques too, such as enrichment, Pakistan has had help and assistance from China. It has concentrated on weapons development, and conducted its first underground nuclear explosion in May 1998.

Conclusion

The history of progress in nuclear energy teaches us many lessons. For the industrialised world, a firm determination, as France had in 1972, resulted in the rapid expansion of nuclear power for electricity production. From technology initially obtained on licence from Westinghouse, France evolved a standardised design and built many reactors, from 600 MW to the latest 1450 MW. About 80 % of the electricity consumption in France is provided by nuclear reactors, and 15 % of the total generated power is exported to neighbouring countries. In all other aspects of the nuclear fuel cycle, from mining to enrichment and reprocessing, France has established commercial viability. It owns uranium mines in other countries, produces low enriched uranium for fuel, with an export market, and has taken a large contract to reprocess irradiated fuel from Japan to recover plutonium and return it to Japan with the residual waste. It has also established competence in fast reactors. The Super Phoenix is an example of a commercial fast reactor of high capacity, over 1000 MW, and it has run successfully. It has had problems not connected to the nuclear aspects, but in the handling of liquid sodium. The capital cost has been high, but is likely to come down with greater standardisation and redesign. In most of the industrialised nations, including Russia, the US and Japan, nuclear energy provides over 20 % of the electricity generated. Many countries in Europe, especially the newly formed countries of the Commonwealth of Independent States (CIS), depend entirely on nuclear power.

It seems therefore that fission power is a viable alternative for the immediate future. But its penetration into the market has taken time. A number of factors have impeded its progress: fear of the unknown, lack of energy supply security through the need to import fuel for the lifetime of the plant, inadequate details and cost for waste and disposal of the waste; plus high capital cost in foreign exchange. In any modern society, a large part of the cost is due to labour, even in hi-tech industries, a cost which is extremely high in industrialised countries. If there is a gradual shift in component manufacture to the developing countries, this may reduce the capital cost to some extent. India has demonstrated this in adequate measure. The cost of nuclear power stations in India is entirely in Indian currency, and is now competitive with even advanced fossil fuel power stations.

The unit size and its compatibility with the existing power grid capacity is another major problem. Fluctuations in voltage and frequency, arising out of mismatch between supply and demand, are a technical problem for nuclear power stations. The Indian experience shows that it could become a major problem in the developing countries. Reactors whose output follow the demand could be an ideal solution, but that is not a priority technology in the developed countries. There the nuclear power stations are more or less base-load stations.

In many countries regulations with respect to licensing procedures, safety standards, and policing of safety measures have created new situations which are peculiar to this industry. The codes and standards generated by international collaborations, through the IAEA (International Atomic Energy Agency), have helped to establish uniform standards irrespective of national standards. This would help in removing the fear from the mind of the general public regarding possible accidents like Chernobyl. But this also calls for a safety culture which unfortunately does not exist in many developing countries. Expertise has to grow in order to ensure safe operations. Manpower training and facilities for on-the-job training in other nuclear power stations in the world could go a long way in enhancing safety. Here again there are no big programmes of international stature, like that of the Atoms for Peace plan. The IAEA could expand its manpower training programme and remove it from the operation of safeguards and sanctions.

To scientists, it seems an enigma, that while the pioneers experimented with a number of reactor systems, it is only the light-water system that has hit the market-place. Equally safe, and technologically better developed systems, like the high-temperature gas-cooled reactor, developed in Germany, US and Japan, have been abandoned as though they are not viable. Perhaps with India and China taking a more aggressive attitude in generating new technology for nuclear power, this limitation may go. The fast reactor is another example wherein, after developing a viable system, the advanced nations have abandoned the idea of commercial exploitation.

It is clear from the above that the role that nuclear energy can play in the future is not entirely dependent on technology but has become political in character, particularly after the introduction of the non-proliferation regime. While everybody in the world would welcome and would like to see a world free from nuclear weapons, they cannot accept an asymmetric situation, wherein a small percentage of the population reserves the right to maintain a nuclear weapons capability. Efforts in universal disarmament have never been very successful, including those of chemical and biological weapons. It is only the technological disparity between nuclear and other areas of warfare, that have led to the belief that restricting technology could ensure capping the capability for nuclear weapons. Even very small technical developments in less powerful countries have led to panic situations. It is necessary to reconcile these inconsistencies and fears for the future.

The question of sovereignty and the ability to think and take decisions for oneself is at the root of democracy. While human rights and respect for democratic values as conceived and propagated by some of the industrialised countries are welcome, it would not be possible to accept sanctions and restraints which almost touch on the sovereignty of nations to develop on their own, whether in technology or in security matters.

Appendix: Milestone Chart and Important Dates[56]

1944-03-12: Dr. Homi Bhabha writes to Sir Dorab Tata Trust to initiate nuclear research
1945-06-01: Tata Institute of Fundamental Research (TIFR) established
1948-08-10: Atomic Energy Commission is set up by Govt. of India.
1954-08-03: Department of Atomic Energy is created by Govt.
1955-04-10: Canada offers assistance to build CIR, an NRX type reactor.
1958-07-28: Decision to build Plutonium Plant is taken.
1959-01-30: Uranium Metal Plant at Trombay produces nuclear grade uranium.
1962-08-09: Nangal Heavy Water Plant starts production.
1969-10-02: TAPS: Tarapur Atomic Power Station starts commercial operation.
1972-05-18: PURNIMA -1, the plutonium fuelled fast critical facility, attains criticality.
1972-08-11: RAPS-1: Rajasthan Atomic Power Station Unit I becomes critical.
1974-05-18: Peaceful Nuclear Explosion experiment at Pokharan.
1985-03-05: Nuclear Waste Immobilisation Plant at Tarapur is commissioned.
1985-10-18: Fast Breeder Test Reactor (FBTR) attains criticality.
1998-05-11: Pokhran-98: Three nuclear devices were tested.
1998-06-21: India and Russia sign Kudankulam agreement for two VVER reactors.

References

P.K. Iyengar, chapter 17 in *Nuclear Power: Policy and Prospects*, P.M.S. Jones (ed.), John Wiley & Sons, 1987.

C.V. Sundaram, L.V. Krishnan and T.S. Iyengar, *Atomic Energy in India, 50 years*, Department of Atomic Energy, C.S. Marg, Mumbai 400 039 India.

Indian Nuclear Society, proceedings of a conference on "The Problems and Prospects for Nuclear Energy in Developing Countries" July 21-25, 1997, Indian Nuclear Society, Mumbai 400 094, India.

"Nuclear Energy for Sustainable Development", International Seminar, Sept. 8-9 1997 Bhabha Atomic Research Centre, Mumbai 400 085 India.

---

[56] This is a shortened version of the milestone chart presented at the Workshop.

Chapter 6 Energy in a Changing World

by
Gert G. Harigel*

Introduction[57]

Previous chapters have discussed the prospects for energy supply and demand world-wide. In this chapter we focus on the special factors that may affect developing and underdeveloped countries in this context and, for illustration, we compare and contrast the scenario in a highly industrialised country - Germany - with that in two major developing countries - China and India.

1. Distribution of electric energy consumption

The countries of the world can be classified, according to their economy and their living standard, into three major groups (see table 1): **(A)** Developed countries, **(B)** Developing countries, and **(C)** Least-developed countries.[58] There is no general agreement which countries belong into which category, due to the vast number of possible criteria that could be applied.

A. Developed Countries

Albania, **Armenia,** Austria, Australia, Azerbaijan, Bahamas, Belarus, **Belgium,** Bermuda, **Bulgaria,** Bhutan, Bosnia-Herzegovina, **Canada,** Croatia, **Czech Republic,** Denmark, Estonia, **Finland, France,** Georgia, **Germany,** Greece, Greenland, Grenada, **Hungary,** Iceland, Italy, Ireland, Israel, **Japan, Kazakstan,** Kyrgyzstan, Latvia, **Lithuania,** Luxembourg, Macedonia, Maldives, Moldova, **Netherlands,** New Zealand, Norway, Poland, Portugal, **Romania, Russia, Slovak Republic, Slovenia, Spain, Sweden, Switzerland,** Tajikistan, Turkey, Turkmenistan, **Ukraine, United Kingdom, United States,** Uzbekistan.

* Gert Harigel is a senior physicist emeritus (CERN, Geneva) and a former visiting professor of physics at Columbia University (New York) and at the University of Hawaii. He is the Secretary of the Geneva International Peace Research Institute (GIPRI) and a Council Member of the International Network for Engineers and Scientists (INES).

[57] This chapter is a shortened version of the paper presented at the Workshop.

[58] SIPRI Yearbook 1995, *World Armament and Disarmament*, Oxford University Press, 1995, pp. 510, 51.

B. Developing Countries

Algeria, Angola, **Argentina**, Bahrain, Barbados, Bahamas, Belize, Bolivia, **Brazil**, Brunei, Cameroon, Cambodia, Chile, **China**, Columbia, Congo, Costa Rica, Cote d'Ivoire, ***Cuba***, Cyprus, Dominica, Dominican Republic, Ecuador, ***Egypt***, Fiji, Gabon, Ghana, Guatemala, Guyana, Honduras, **India**, Indonesia, ***Iran***, Iraq, Israel, Jamaica, Jordan, Kenya, ***North Korea***, **South Korea**, Kuwait, Lebanon, Libya, Madagascar, Malaysia, Mali, Marshall Islands, Mauritius, **Mexico**, Fed. States of Micronesia, Mongolia, Morocco, Namibia, Nicaragua, Oman, **Pakistan**, Panama, Papua New Guinea, Paraguay, Peru, Philippines, Qatar, St Vincent & the Grenadines, El Salvador, Saudi Arabia, Senegal, Seychelles, Singapore, Solomon Islands, **South Africa**, Sri Lanka, Surinam, Swaziland, Syria, **Taiwan**, Thailand, Tonga, Trinidad & Tobago, Tunisia, Tuvalu, United Arab Emirates, Uruguay, Venezuela, Vietnam, Zaire, Zambia, Zimbabwe.

C. Least Developed Countries

Afghanistan, Bangladesh, Benin, Botswana, Burkina Faso, Burundi, Cape Verde, Central African Republic, Chad, Comoros, Djibouti, Equatorial Guinea, Eritrea, Ethiopia, Gambia, Guinea, Guinea-Bissau, Haiti, Kiribati, Laos, Lesotho, Liberia, Malawi, Mali, Mauritania, Mozambique, Mianmar, Nepal, Niger, Rwanda, Samoa, Sierra Leone, Somalia, Sudan, Tanzania, Togo, Uganda, Vanuatu, Yemen, North Yemen, South Yemen.

Bold & underlined: Countries with nuclear power stations. ***Bold, italic & underlined***: Countries with nuclear power stations under construction. ***Bold & italic***: Countries with nuclear power stations in the planning stage.

Tab. 1. Classification of countries according to their stage of economy and living standards.

The consumption of electric energy is not evenly distributed among countries, neither in per capita consumption of electricity, nor in the fraction of energy for a given country that is electric (see table 2). Tanzania and Bangladesh use less than 100 kWh per person per year[59], while in China the figure is about 1 000; in the Republic of Korea 5 000 and in Japan 7 500. Among the globally greatest consumers are the US with about 12 000, Sweden with 15 000 and hydropower-rich Norway with the highest figure: 25 000.[60]

---

[59] 1 kWh/y = 0.114 W: see the Appendix of this book.

[60] Hans Blix: "The global need for nuclear power in the 21st century" at the *Tokyo Asia Atomic Energy Forum*, March 6-7, 1998, *For the Renaissance of Nuclear Power in Asia*, compiled by Kumao Kaneko, p. 5.

2. Energy consumption patterns

To understand more clearly the different energy consumption patterns for the various stages of development among nations, the energy situation will be evaluated for a few selected countries and regions from each of these groups. This will establish a framework for analysis, make some comparisons possible, and permit some preliminary conclusions to be drawn.

	Countries		
	Developed	**Developing**	**Least Developed**
Number of countries	57	87	41
Population (million)	1 375.2	3 928.1	515.0
Percentage of world population (%)	23.6	67.5	8.8
Per capita energy consumption (t BCU / y) [(a)]	5.633	1.009	0.170
per capita coal (t BCU / y)	1.880	0.480	0.002
per capita gas (t BCU / y)	1.591	0.130	0.004
per capita oil (t BCU / y)	1.877	0.344	0.031
per capita hydro (t BCU / y)	0.285	0.055	0.133
Per capita electricity consumption (kWh / y)	6 360	830	54
Per capita electricity consumption (t BCU /y)	0.782	0.102	0.007
Number of countries with nuclear power	23	9	0
Number of nuclear reactors	395	38	0
Nuclear electric power (GW(e))	323	22.5	0
Nuclear / total electric energy (%)	23.8	4.4	0
Percentage of world nuclear electricity (%)	94	6	0

(a) 1 t BCU/y corresponds roughly to 1 kW.

Tab. 2. Distribution of Electric Energy Consumption.
Source: 1998 Britannica Book of the Year.[61]

3. Energy consumption patterns for Germany

The energy requirements of the industrialised country, Germany, will help in an analysis of whether nuclear energy is absolutely needed, just a convenience, or could be completely abandoned. Germany has a high living standard and a stable population and

[61] *1998 Britannica Book of the Year*, Encyclopædia Britannica, London, 1998, pp. 756-761 and pp. 820-825. Data on population, energy consumption per capita (coal, gas, oil, hydro-electric, without wood), electric energy consumption per capita, and usage of nuclear electric energy - all taken from the *1998 Britannica Book of the Year* (for 217 countries) - are grouped according to the countries' stage of development (see table 1). The energy data are mainly for 1994.

comprehensive data are available for it, allowing a thorough analysis.

The total energy needed by mankind for various essential or non-vital purposes can be conveniently subdivided into energy in the form of (i) food, (ii) fuel, (iii) electricity, (iv) low-temperature heat for buildings and warm water, and (v) high temperature process heating.[62]

A food requirement of 3 500 kilocalories per person per day (for an explanation of units see the Appendix), is the amount needed to keep the population healthy; this amounts to 1.4 EJ per year. The actual agricultural bio-mass from 47 % of the country's surface produces 1.6 EJ / year, resulting in a small overproduction. In 1995 12 % of the grain area in the European Union had been held out of production, but this was reduced to 5 % in 1997, including the area used in Germany.[63] This action was a consequence of a decrease of world-wide carry-over food stocks, approaching a food stockpile sufficient for only fifty days.[64] More land will be needed for agricultural use, making less available for other purposes.

Fuel requirement is 60 million tons of gasoline, which corresponds to 2.6 EJ per year. It is covered by 70 million tons of petroleum per year. Germany will never be independent of oil imports, since it can never produce enough bio-mass for conversion into fuel. The amount of bio-mass equivalent to this oil demand is 5 EJ/year; this would have to be harvested from 1.5 times the surface of Germany. The automobile fleet is approaching a saturation value in Germany, as in most industrial countries; however the world-wide automobile fleet is expected to double within the next 25 years and will consume one third of all oil production.

Electricity required is 530 TWh per year. This quantity could come from plants burning 170 million tons of bituminous coal. It could also come from 12 000 tons of natural uranium in nuclear power plants. It could *not* come *solely* from hydroelectric power stations, since the equivalent of 40 fully-exploited Rhine rivers would then be needed. If solar photo-voltaic cells were to be used, 7 000 km^2 of solar cells would be required, combined with the storage capacity of eight billion tons of batteries (covering the area of the State of North-Rhine - Westphalia = 32 000 km^2), or stored in storage dams (each with the physical characteristics of the *Edertalsperre*) covering a total area the size of former West Germany. Only the first two energy options, coal and nuclear, can provide a major fraction of Germany's electric-energy needs, although hydro- and solar-electric production methods should be exploited as much as possible to feed some supplementary energy into the electric grid.

Heating of buildings and hot water (less than 100 C) consume 3 EJ per year. This requirement is presently covered by a combination of burning bituminous coal, petroleum and natural gas. To provide this energy by burning wood, would require an annual consumption of 450 million m^3 of wood. This would have to be harvested from a surface

---

[62] Klaus Heinloth, *Die Energiefrage, Bedarf und Potentiale, Nutzen, Risiken und Kosten (The Energy Challenge, Requirements and Potentials, Yields, Risks and Costs)* Vieweg, Handbuch Umweltwissenschaften, Wiesbaden, 1997, pp. 486-9.

[63] Lester L. Brown, Michel Renner and Christopher Flavin, *Vital Signs 1997, The Environmental Trends that are Shaping our Future*, W.W. Norton for the Worldwatch Institute, New York, 1997, p. 40.

[64] Brown, *et al.*, *ibid.*, p. 35.

2.5 times the area of Germany. A hypothetical case can be made for combining 2 000 km^2 of flat solar collectors with 20 km^3 of hot water storage.

High temperature process heat, often forgotten in discussions about energy, requires the considerable amount of 1.8 EJ per year, which is larger than the demand for food energy. It could be hypothetically covered by 80 million tons charcoal per year, or by burning 300 million m^3 of wood per year, which corresponds to the wood production of 1.5 times the surface of Germany. If this demand were to be covered by the use of electrical energy, 1.9 EJ would be required. That would double the demand for electric energy.

The data indicate that the import of fossil products remains a necessity for Germany since no (complete) replacement for it from other sources can be envisaged under any reasonable assumptions. Renewable energy can not be expected to satisfy more than a few percent of the energy demand. Alternative energies require a lot of land surface. Conservation in the form of better insulated buildings, reducing the temperatures in living and working spaces, disconnecting unnecessary electrical equipment, and designing more efficient conversion engines, could become a high priority to at least slow the increase in energy demand. However, nuclear energy is still needed at the present level to supply about 33 % of Germany's energy needs for many years to come. The number of power stations might even profitably be increased.

4. Energy for big, rapidly developing countries

China and India will illustrate the energy consumption and production patterns of rapidly developing countries and are compared with two highly industrialised countries, Germany and the United States. In considering data from 1996, it is necessary to keep in mind the relative populations of China, India, Germany, and the United States, namely 1219, 953, 82, and 265 million, respectively (see table 3).[65]

Both of these large developing countries face a population explosion, rapid industrialisation, and a substantial loss of arable land. China had a crop land loss of 3 percent from 1986 to 1992, through growth of cities, the building of roads, and the expansion of deserts.[66] It also experienced a decrease in forest area of somewhat less than one percent per year.

China's aggravating energy problem originates from its population explosion and the concentration of most of its mega-cities and major industry in coastal areas. China's coal reserves are in the northeast and central northern parts, far away from the place where energy is needed. Transport is already inadequate and a crushing burden for the railways to move sufficient quantities of coal (see Chapter 5, by Iyengar).[67] Hydroelectric

---

[65] *1997 Britannica Book of the Year*, Encyclopædia Britannica, London, 1997, pp. 612, 627, 741 and *1998 Britannica Book of the Year, op. cit.*, p. 577.

[66] Lester L. Brown, *Who Will Feed China, Wake-up Call for a Small Planet*, The Worldwatch Environmental Alert Series, W.W. Norton, New York, 1995, pp. 54-65.

[67] Blix, *The global need for nuclear power in the 21st century, op. cit.*, p. 16.

power is in the south-west, again relatively far from places where it is mostly needed.[68] It is a well-known fact that any transport of electricity over large distances is coupled with considerable losses.

The situation of dwindling aquifers aggravates the situation, both for drinking water supply, and water for irrigation. Throughout the world, hundreds of desalination units, producing from a few thousand to 37 900 000 litres (10 000 000 gallons) or more per day, already are in operation. In general, the desalination plants in production are where high-cost desalinated water can be afforded. A population usually can afford to pay about 10 times as much for water for domestic purposes as it does for agricultural water. Proposals for large-scale *nuclear* desalination facilities promise to lower the cost of desalinated water at the desalination sites to a level that most industries and a few agricultural enterprises can afford. This can be considered an additional argument in favour of having nuclear power plants close to China's coast line.

Similar conditions and requirements exist for the Indian subcontinent. The concentration of rapidly growing mega-cities require large amounts of electricity, which can be provided only on a tiny scale by renewable energies. Again, nuclear power might become the preferred option (see Chapter 5, by Iyengar).

In addition to these detrimental developments there is a high degree of pollution. As the standard of living increases in China, there is an increased food consumption, climbing up the food ladder, leading possibly to a severe food shortage in the future. Its grain lands dropped over the last forty years by 50 % to 0.07 hectares per person and may fall a further 50 % by the year 2030. While the grain production per person is still stable in India at 180 kg per person per year, it appears already to be dropping slightly for China from the present 290 kg per person per year.[69]

Both countries face the dilemma of where to find increasing energy resources, particularly electric energy. These energy resources are required to satisfy the demand for better transportation and growing industries. They could also be useful in the production of drinking water and water for irrigation by desalination, which might become necessary due to a dramatic drop of aquifer levels.

The growth of the automobile fleet needs also to be accommodated. While this fleet is presently still modest in size, increasing from about 1 million in 1990 to the present 2.7 million, the government predicts a growth by almost an order of magnitude by the year 2010 to reach about 22 million cars.[70]

China has 10 % of the World's fossil fuel resources, including 12 % of the coal, 2.4 % of the oil, and 1 % of the natural gas. China's coal consumption is four times larger than that of Germany, has risen by 50 % during the last decade, and was climbing by 4.5 % in 1996.[71] Even though China's electricity production is presently only 30 % more than that of Germany, the use of coal on a large scale in power stations, without appropriate removal of hazardous particulate matter, has created a severe pollution problem. Nuclear power plants produce only 1.5 % of electric energy, with three nuclear

---

[68] *Encyclopædia Britannica*, 15th Edition, 1988, Vol. 16, p. 54.

[69] Brown, *Who Will Feed China, op. cit.*, p. 29.

[70] Brown, *Who Will Feed China, op. cit.*, p. 58.

[71] Brown, *et al.*, *Vital Signs 1997, op. cit.*, p. 46.

reactors in operation, generating 2 100 MW. However, nuclear electrical energy is intended to go up to 20 000 MW by the year 2010, using two nuclear reactors that are under construction and a further four that are in the planning stage. China supposedly plans to have about 240 nuclear power reactors by the year 2050.[72] Even this number of reactors would satisfy only 13 % of China's electric needs by then.

	Produced	**Consumed**
China		
Electricity	928 083 000 000 kWh	926 037 000 000 kWh
Coal	1 239 902 000 tons	1 231 928 000 tons
Crude petroleum	1 069 320 000 barrels	1 024 375 000 barrels
Petroleum products	106 629 000 tons	114 972 000 tons
Natural gas	17 540 000 000 m^3	17 540 000 000 m^3
India		
Electricity	351 000 000 000 kWh	385 902 000 000 kWh
Coal	273 859 000 tons	284 497 000 tons
Crude petroleum	244 743 000 barrels	434 149 000 barrels
Petroleum products	43 575 000 tons	56 722 000 tons
Natural gas	17 638 000 000 m^3	17 638 000 000 m^3
Germany		
Electricity	528 221 000 000 kWh	530 558 000 000 kWh
Hard coal	57 623 000 tons	66 255 000 tons
Lignite	207 077 000 tons	209 308 000 tons
Crude petroleum	21 535 000 barrels	793 500 000 barrels
Petroleum products	99 578 000 tons	113 839 000 tons
Natural gas	20 904 000 000 m^3	92 770 000 000 m^3
USA		
Electricity	3 268 250 000 000 kWh	3 312 888 000 000 kWh
Coal	937 580 000 tons	843 873 000 tons
Crude petroleum	2 464 000 000 barrels	5 024 000 000 barrels
Petroleum products	704 201 000 tons	737 681 000 tons
Natural gas	530 014 000 000 m^3	592 209 000 000 m^3

Table 3. Energy production and Consumption (1994).
Source: 1997 and 1998 Britannica Book of the Year.[73]

In India, electricity production is two-thirds that of Germany, with 1.5 % of it

---

[72] See Chapter 11, by Garwin: quote from CISAC (US Committee on International Security and Arms Control, National Academy of Sciences), Beijing discussions, 1997.

[73] *1997 Britannica Book of the Year*, *op. cit.*, pp. 612, 627, 741, and *1998 Britannica Book of the Year*, *op. cit.*, p. 577.

nuclear; whilst total coal consumption equals that of Germany. Since India possesses only limited fossil-fuel resources, and faces a huge population increase, it may have to go more toward the production of nuclear electric energy.

China and India, representing one third of human population, have made, until now, little use of the nuclear sector, which averages world-wide at 17 %. Table 4 gives the present electricity production according to production means in China and India, as well in two other, adjoining South and East Asian countries. Some increase in the use of hydroelectric power stations may be possible in principle. However, such major projects may lead to international political struggles over flood control, as well as over the availability of drinking and irrigation water downstream along the big rivers. The "Three Gorges Dam" on the Yangtze river is planned to produce 18 200 MW electric energy, which corresponds to 18 1-GW(e) nuclear power stations. It is scheduled to be operational by the year 2009. It will contain a 600 km long reservoir, and will take out 240 000 acres (970 km^2) of cropland. Out of the 9 573 million km^2 of China's land surface only ten percent are considered to be cultivable, i.e. the dam will reduce by 0.1% the fertile crop land. Arguments for building the dam, apart from "clean" electricity production, are the control of floods and protection of humans from their devastation. Among other detrimental effects there is the non-negligible danger of damage by earthquakes and its catastrophic consequences.

	Hydroelectric	**Main River(s)**	**Fossil**	**Nuclear**
Pakistan	34.0 %	Indus	65.1 %	0.9 %
India	18.5 %	Ganges	80.0 %	1.5 %
China	18.1 %	Yangtze	80.4 %	1.5 %
Bangladesh	6.0 %	Ganges, Brahmaputra	94.0 %	0 %

Tab. 4. Relative share of electricity production of different origin in selected South and East Asian countries.[74]

China and India have put a considerable effort in the improvement of (scientific) education, which has resulted in a remarkable number of high quality scientists. In both countries there are probably enough qualified scientists and technicians to operate nuclear power plants in a safe way. The educational level is superior to that in most of the other developing countries (see Chapter 5, by Iyengar).

Evidently, solar cells, windmills and bio-mass will not be capable of providing to-day's or tomorrow's urbanised modern societies - the Shanghais and Calcuttas - with the huge amounts of energy, particularly electricity, which they will require. To have the equivalent of a 1 000 MW(e) nuclear or coal plant, using solar cells, an area of more than 20 km^2 would have to be covered by such cells. If one has to rely on windmills, one would need wind farms covering more than 50 km^2. And if one had to rely on bio-mass

[74] *Britannica Book of the Year 1998, op. cit.*, pp. 820-825.

(wood) one would need plantations of more than 4 000 km^2 of land.[75]

A comparison of various sectors of energy production and consumption between China and India on the one hand, and the United States on the other, demonstrates huge differences in absolute values and the way they may develop (see table 3).

5. Energy for least-developed countries

The category of least-developed countries includes all Sub-Saharan countries close to the equator. That implies that there is less need for energy for heating purposes, but electric energy for cooling is highly desirable. Basically for these countries the competition is between the use of scarce renewable wood resources for fulfilling minimal food and heat and transport requirements - and the need to use these for improving industry and supporting infrastructures. For example, increased use of scarce firewood for cooking is leading to the over exploitation of land, turning it into desert.

Could these countries really utilise substantially more electricity in an effective way? If so, huge investments would be needed for a future electricity grid and its maintenance. If much electric energy could indeed make a major improvement for the least-developed countries, is nuclear electric power the right solution? There is a doubt that such countries, having a low level of technical and scientific education, can be trusted with the operation and maintenance of nuclear reactors. Might solar electric energy, produced in small local installations or solar operated stoves, be a solution to the need? If these countries do need more electric energy, would it not be better to transport it to them from developing or developed countries?

World-wide, at the end of 1996, there were 433 nuclear reactors operating, producing 345.5 gigawatt of electric power (GW(e)). Another 45 reactors were under construction for a further net output of 38.7 GW(e). The net annual generation of the existing reactors is 2 228 TW hours or 254 GW years, corresponding to 17 percent of the world-wide electric energy production. Table 5 gives, world-wide, the data for all countries possessing nuclear power reactors.[76] Among these are, as indicated, both developed and developing countries. None of the least developed countries have nuclear power, as previously shown in table 1.

[75] Blix, *The global need for nuclear power in the 21st century, op. cit.*, p. 20.

[76] American Physical Society (APS), available at http:// aps.org/public_affairs/ popa/popaii-2.html.

Country	Operating end 1996		Under Construction end 1996		Net Generation 1995		Percent Nuclear
	Number	Net (GWe)	Number	Net (GWe)	TWh	GWy	1995
US	109	100.5			673	76.9	22
France	55	57.4	4	5.8	359	40.9	76
Japan	51	41.0	5	5.7	287	32.8	33
Germany	20	22.2			154	17.6	29
Russia	26	19.8			99	11.3	12
Canada	22	15.4			92	10.5	17
UK	35	12.7			78	8.9	25
Sweden	12	10.1			67	7.6	47
Ukraine	14	12.1	4	3.8	66	7.5	38
South Korea	11	9.1	9	7.6	64	7.3	36
Spain	9	7.2			53	6.1	34
Belgium	7	5.5			39	4.5	56
Taiwan	6	4.9	2	2.7	34	3.9	29
Switzerland	5	3.1			24	2.7	40
Finland	4	2.3			18	2.1	30
Bulgaria	6	3.4			17	2.0	46
Hungary	4	1.7			13	1.5	42
China	3	2.1	6	4.6	12	1.4	1
Czech Republic	4	1.6	2	1.8	12	1.4	20
Slovak Republic	4	1.6	2	0.8	11	1.3	44
South Africa	2	1.8			11	1.3	6
Lithuania	2	2.8			10.6	1.2	86
Mexico	2	1.3			8	1.0	6
Argentina	2	0.9	1	0.7	7	0.8	12
India	10	1.7	6	1.7	6	0.7	5
Slovenia	1	0.6			5	0.5	40
Netherlands	2	0.5			4	0.4	5
Brazil	1	0.6	2	2.5	2.5	0.3	1
Pakistan	1	0.1	1	0.3	0.5	0.1	1
Kazakstan	1	0.1			0.1	0	0
Romania	1	0.7	1	0.6	0	0	0
Armenia	1	0.4			0	0	0
World Total	433	345.5	45	38.7	2228	254	17

Tab. 5. World nuclear status: number of reactor units and total capacity, operating and under construction (end 1996); nuclear generation, and nuclear share of the electricity generated by utilities (1995). *Source: American Physical Society (APS).*

Conclusions

The world will undoubtedly face a severe energy crisis, starting early in the coming decades and reaching its climax in the second half of the twenty first century. Energy supplies are intrinsically related to the well-being and survival of humans all over the globe, and with their living together in peace. Therefore a major challenge to all countries is to provide sufficient energy resources to their people. The question here is how much nuclear electric energy is needed in the mix of energy for countries at different levels of development. Are nuclear power reactors an appropriate solution for not only developing countries but for least-developed countries as well? Is a transfer of nuclear-reactor technology to least-developed countries appropriate?

The most important social challenge is to guarantee the production of enough food and the provision of safe drinking water in order to minimise the number of deaths caused by starvation and disease. At present, 18 million people per year are victims of these food and clean-water shortages, caused not primarily by production limitations, but rather by inadequate distribution of food due to financial considerations, the unavailability of safe water, and inadequate medical care. This provision of food and water is not directly linked to energy supplies. In fact, the desire for more energy can interfere with satisfying these basic needs. For example, it seems inadvisable to use agricultural land for the production of bio-mass fuels for transportation and electric energy.

The second task, not less challenging, is to narrow the gap in standard of living between rich and poor countries. This implies that sufficient energy has to be available in all countries to encourage structural improvements, technological developments in industry and increases in living standards. Electrical energy can assist in technological development, as well as having the advantage of convenience in all fields of life, be it in powering engines for mass transportation, machinery in factories, illumination, heating, household equipment, or information systems (radio, television, telephone). The latter, for example, offer direct improvements in both social structures and in living standards through help with better education and improved social interactions. Thus electric energy is vital for closing the gap between developed and least-developed countries.

However, electrical energy cannot be considered in isolation from broader energy demands and supplies. It has to be seen in the context of the variety of total energy requirements. Each nation and region has to make its own decisions as to what fraction of its total energy should be contributed as electricity. Within this context of need for electrical energy, the next decision must be how much of that electrical energy should be provided by nuclear power. Despite a reasonably good performance by nuclear power stations, there is a strong anti-nuclear movement in several, mainly industrialised countries, which can - in the author's opinion - be attributed in a large degree to insufficient knowledge about basic physics and comparative risk factors. However, *none* of the production mechanisms are free of adverse side effects and *none* is absolutely safe. The data from this and preceding chapters suggest that the world will have to live with nuclear energy for the near and probably more distant future, or else may have to drastically reduce the consumption of electricity.

Which countries would find nuclear electrical power most suitable? Future building of nuclear reactors in *developing countries,* such as China and India, can be justified on ecological grounds; and may well be recommended, provided some pre-conditions can be fulfilled.

(i) Only modern, safe and cost-effective nuclear reactors should be installed. For instance, the heavy-water reactor of the CANDU type (Canadian Deuterium Uranium reactor) has a high initial construction cost, but provides the important advantage of using natural uranium. This avoids the costly enrichment process and the dependence on delivery of enriched uranium-235 from developed countries. The absence of enrichment plants in the developing countries would reduce the danger of clandestine accumulation of weapon-grade material.

(ii) Accident and proliferation risks must be minimised by using operators who are well trained and supervised, perhaps even employed by the International Atomic Energy Agency in Vienna (IAEA). A second reactor accident like that at Chernobyl, whether caused by human operator mistake or equipment failure, would deal a deathblow to production of nuclear electric power all over the world.

(iii) The radioactive waste problem must be addressed at an early stage. The situation for *least-developed countries* is far more complex. For these countries there is also the need for an increase in electricity supply. But for them there is the additional problem that the effectiveness of a nuclear-electric system must first be established. It will be much more difficult to afford those modern safe and cost-effective reactors. Accident risks may be higher because of a weak infrastructure. And the radioactive waste problem can not be solved internally. Given the weak infrastructure of most of these least-developed countries, it might be better to start with imported electricity, wherever this is possible.

In general, decisions about nuclear electric power must be made in the very widest context of an over-all energy policy. Long-term political and economic decisions have to be made, the sooner the better. Energy priorities must be defined to avoid wars over the possession of energy resources, still abundant in some developing countries, such as access to oil wells. The production of any form of energy should not be achieved by sacrificing food production from arable land for secondary amenities; not in the developed, developing, or least-developed world.

The pros and cons of renewable energies, many easily being overlooked, should be carefully evaluated and their impact on the environment and bio-diversity studied. Only then can the need for nuclear electric energy be completely evaluated. A comprehensive analysis of the economic, social, and environmental energy situation has to be made, before it is possible to judge the desirability of transferring nuclear technology to least-developed countries. Use of nuclear energy would probably prevent - or at least slow down - further, drastic depletion of vital resources and destruction of the environment in less developed countries. A closer collaboration between scientists and politicians on the energy subject is indispensable! There must be continued effort in the fields of education and public information, since benefits and hazards still seem almost mystic to many and, therefore, particularly frightening. This is a tremendous challenge for the scientific and technical community.

Chapter 7 Safety of Nuclear Power - Some Observations

by
Jack Harris*

Introduction

Preparation of this chapter has been greatly helped by involvement in four particular events: (a) Immediately following the Windscale accident in October 1957, I spent a month at the site rescuing graphite specimens from the stricken reactor and its twin, and carrying out experiments with these samples; (b) Shortly after the Three Mile Island (TMI) accident, in 1979, I was working at Oak Ridge and had discussions with a number of scientists who had been involved in assessing the causes of the incident; (c) I had the opportunity in 1995 to visit the Chernobyl site and was taken on a tour of inspection inside the sarcophagus, where I viewed the control room of the No 4 reactor where much of the initial drama had taken place. I was able to talk to some of those who had been involved in the accident and to interview the Minister in charge of nuclear power in Kiev. My particular interest has been to study to what extent the operators were responsible for the three accidents and to what degree the operators were made scapegoats for shortcomings much higher up in the command structure; and finally (d) in 1998, I participated in a European Commission seminar on "Nuclear in a Changing World", where I was fortunate enough to hear L. Hogberg's lecture "Safety Issues in Nuclear Power", and to read his paper, from which I have borrowed freely.[77] I have also been very impressed by the evident wisdom (and elegant prose) of the publications of the IAEA (International Atomic Energy Agency) International Nuclear Safety Advisory Group (INSAG). To illustrate, here is INSAG's definition of the "Defence in Depth" concept:

"Defence in depth is generally structured in five levels. Should one level fail, the subsequent level comes into play. The objective of the first level of protection is the prevention of abnormal operation and system failures. If the first level fails, abnormal operation is controlled or failures are detected by the second level of protection. Should the second level fail, the third level ensures that safety functions are further performed by activating specific safety systems and other safety features. Should the third level fail, the fourth level limits accident progression through accident management, so as to prevent or mitigate severe accident conditions with external releases of radioactive materials. The last objective (fifth level of protection) is the mitigation of the radiological consequences of significant external releases through the off-site emergency response ...

* Jack Harris is Editor of Interdisciplinary Science Reviews and formerly Senior Section Head, Berkeley Nuclear Laboratories, UK.

[77] L. Hogberg, "Safety Issues in Nuclear Power", European Commission Seminar: *Nuclear in a Changing World*, Brussels, 14-15 October 1998.

... The general objective of defence in depth is to ensure that a single failure, whether equipment failure or human failure, at one level of defence, and even combinations of failures at more than one level of defence, would not propagate to jeopardise defence in depth at subsequent levels. The independence of different levels of defence is a key element in meeting this objective."

1. Relevance of "Defence in Depth" to the Three Mile Island Accident

A key feature of defence in depth is the existence of multiple barriers against release of radioactivity to the environment. Such barriers are:

(1) the retentive capability of the ceramic fuel pellets;
(2) the fuel cladding;
(3) the pressure vessel, and
(4) the containment structure.

In the case of TMI, all these barriers were in place though it was the final one, the containment, which saved the day. A substantial fraction of the core content of radioactive nuclides was released into the containment building, but less than 10 % of the noble gases and extremely small fractions of other nuclides, such as iodine and caesium, were released to the environment. The total release from the accident has been estimated to correspond to about a decade of releases from normal operation, and can be regarded as negligible. Of course, the local population was unaware that the level of leakage *was* very low and near panic ensued. This emphasises the need for good information flow and for the immediate distribution of iodine tablets; and also for well-rehearsed procedures for the public to take cover inside homes and other buildings, and eventually for controlled evacuation where necessary.

The Chernobyl No 4 so-called "RBMK" reactor did not have the benefit of a strong secondary containment, and, even if it had, it is not certain that it could have withstood the uncontrolled release of *nuclear* energy which was the initiating event of the accident. As a result, all the noble gases and between 20 % and 60 % of the volatile radionuclides, such as iodine and caesium, were released to the environment. Following the accident there was a major review of the safety of the remaining operating RBMKs. In particular, operation and reactivity characteristics, and reactivity control, have been improved and a greater emphasis has been placed on training of operators. Improvements were also made to the operating so-called "VVER" reactors (Russia's version of the Pressurised Water Reactor, PWR), though even with the most modern design of this type there remains some concern with the capability of the containment to cope with severe accident conditions.

2. History and Future of Water-cooled Reactors

While it is of course essential that the RBMK reactors operate without major incident during their remaining lives, it is inconceivable that anyone would wish to design and construct entirely new reactors of this type. In contrast, in the rather unlikely event of the nuclear construction industry picking up in the near future, it is almost certain that any new reactor ordered will be of the light water reactor (LWR) type (probably PWRs) or possibly a Canadian Deuterium reactor (CANDU). To face up to this possibility, it is useful to look at the safety record of water-cooled and water-moderated reactors. By the end of 1999, there will have accumulated close to 10 000 reactor-years of operation of commercial nuclear plants, cooled and moderated with light or heavy water. If, as seems probable, no serious accident occurs before then, throughout this accumulated experience of civil nuclear power, there will have been only one major incident with a commercial light water reactor - the accident at Three Mile Island.

Bearing in mind also that although the TMI core was severely damaged there were no off-site effects, then the record of just one severe core-damage incident in 10,000 reactor-years operation can be regarded with a measure of satisfaction, indeed it will precisely conform with the INSAG target for existing stations. Considering also that, following the TMI incident, significant improvements were made to design and operating conditions of water reactors generally, then one could perhaps look forward to the achievement in the future of a new target of one serious incident per 100 000 reactor-years. It can be concluded that, should an expansion in nuclear power be contemplated in the near future, then efficient and safe water reactor designs exist to meet the challenge.

Nevertheless, for the longer term, if only to regain the confidence of the public, even higher standards may be required, hence the following paragraphs.

3. Water Reactor with Passive Safety Features

The Electrical Power Research Institute in the US has defined the principal safety characteristics of water reactors with passive safety features:

- Completely passive shutdown and cooling systems;
- No external electrical power needed for safety functions;
- Containment function so reliable and effective that early public action need not be required in the event of an accident.

As early as 1992, INSAG reported that passively-safe reactors were under active consideration, mostly with power levels in the range 600 to 800 MW(e).

Even smaller reactors of advanced concept have been designed. The Safe Integral Reactor (SIR) was to be a 320 MW(e) reactor which has all the major components housed in a single large vessel. The Modular High Temperature Gas Cooled Reactor (MHTGC) was to have been a 350 MW(th) graphite moderated reactor in which, instead of having a huge confinement structure, the fuel and resultant fission products would be confined by an impervious coating on the particles of fuel. The US Integral Fast Reactor (IFR) was to have used metallic uranium or plutonium as fuel; and finally, the Process Inherent Ultimate Safety (PIUS) design sought to provide core safety by means of inherent features of thermal hydraulics and gravity in producing natural circulation.

There are no shortage of ideas, it is financial support and a possible market which is lacking. The concentration on small outputs may be out-of-date as the states likely to expand their nuclear power requirements - China, India and South American countries - are moving towards having grid distribution systems of substantial size.

Research and development relating to passive reactors and future designs generally will not provide a return for at least 20 years, so finance from governments and international collaboration is essential if progress is to be made. The strong movement internationally to "liberalise" (i.e. privatise) utilities, which started largely in the UK, must be considered a negative factor here. The advantages offered by the fast reactor should not be overlooked. It is being felt increasingly that sodium cooling is too hazardous (and difficult to sell to the public) so attention is being focused on gas cooled, and lead cooled, fast reactors.

Of the advanced systems listed above, considerable attention is being paid to fuel systems such as are used in the High Temperature Gas-cooled Reactors, i.e. where the fuel is encapsulated inside small graphite spheres. It has been proposed that, were the impermeable graphite coating to be replaced by suitable oxides (alumina, zirconia, magnesia etc.), then such pellets could be used in LWRs, thereby enhancing safety and allowing the achievement of extremely high burn-ups. Where extensive use of Mixed Oxide (MOX) fuel is being contemplated, consideration is being given to removing the problem of generation of plutonium by replacing uranium oxide as the fuel matrix material with thorium oxide. This will result in the generation of uranium-233 which is of course fissile, but which, unlike plutonium, can be "denatured" by additions of uranium-238. Plutonium/thorium MOX would also generate a lower radiotoxicity burden than plutonium/uranium MOX (see Chapter 8, by Schapira).

To end on a more topical note, it seems unlikely that new nuclear power stations will be economically viable for some considerable time. However, as much of the cost of nuclear electricity relates to high capital/construction charges, nuclear power does become highly competitive once these charges are written off, i.e. there is a strong incentive to extend the life of existing nuclear stations. As the great rash of reactor construction took place in the late 1960s and 1970s, many operating stations have already exceeded half their design lifetimes, so the issue is an urgent one. This raises interesting questions on the safety of ageing stations. A relevant issue is: should the same safety standards be adopted for extensions as apply to new stations? Such subjects as the gradual accumulation of irradiation damage to steel pressure vessels come to the fore. The enormous run-down in research activity by privatised utilities becomes a matter for concern. There are also incentives to increase the burn-up of the fuel: targets of 60 MWd/kg for the PWR are being discussed. Of course, with increasing irradiation, the pressure of fission gases inside the zircaloy pins increases, with obvious safety implications. One solution here may be to provide greater safety by manufacturing oxide-coated fuel particles to be placed inside standard pins, as discussed earlier.

Chapter 8 The Nature and Management of Nuclear Wastes

by
Jean-Paul Schapira*

Introduction

The nuclear industry is today facing many challenges in its effort to maintain the nuclear energy option open for the next few decades. Economic viability in the new context of electricity market liberalisation is one of these challenges. Another one is social acceptability, largely determined by the perceived level of reactor safety and proliferation issues. On the other hand, management of nuclear wastes, produced in nuclear reactors and in fuel cycle operations, either for civilian or for military use, has become a very strong environmental issue among the public.

The waste issue is connected essentially to two achievements:

a) the correction of detrimental situations arising from past practices in nuclear activities related to weapons plutonium use, uranium mining, spent fuel reprocessing or others;

b) the demonstration that long-lived nuclear wastes produced inside reactors, mostly commercial, can be dealt with in such a way as to protect future generations and the environment.

The cleaning up of early nuclear weapon plants, the conditioning of secondary wastes which are left in bulk, the rehabilitation of mining sites, for example to prevent radon emission, are increasingly prominent issues. Nevertheless, in some cases the nuclear industry tends to postpone the implementation of such measures for technical or financial reasons. However, the recent 17 billion dollar contract between the US Department Of Energy (DOE) and British Nuclear Fuels Limited (BNFL) for cleaning the various contaminated military sites in the US is worth mentioning.

---

* Jean-Paul Schapira is Director of Research at the French CNRS (*Centre National de la Recherche Scientifique*), *Institut de Physique Nucléaire*, Orsay. He is a former member of the French Cabinet of the Ministry of Environment.

1. General Background on Back-end Nuclear Wastes

Nucleus	Half Life (years, days)	Dose factor a) (Sv/Bq)	Mass b) (g/tHM)	Mass c) (kg/TWh)	Mass d) (kg/year)
Uranium:					
^{235}U	$7.03\ 10^{8}$ y	$4.6\ 10^{-8}$	$10.270\ 10^{3}$	38.9	238.5
^{238}U	$4.46\ 10^{9}$ y	$4.4\ 10^{-8}$	$940.6\ 10^{3}$	3563	21 948.3
Total Uranium			$955.4\ 10^{3}$	3625	22 228.5
Plutonium:					
^{238}Pu	87.7 y	$4.9\ 10^{-8}$	$0.176\ 10^{3}$	0.67	4.1
^{239}Pu	24110 y	$2.5\ 10^{-7}$	$5.673\ 10^{3}$	21.46	131.6
^{240}Pu	550 y	$2.5\ 10^{-7}$	$2.214\ 10^{3}$	8.39	51.4
^{241}Pu	14.4 y	$4.7\ 10^{-9}$	$1.187\ 10^{3}$	4.50	27.6
^{242}Pu	$3.7\ 10^{5}$ y	$2.4\ 10^{-7}$	$0.490\ 10^{3}$	1.86	11.4
Total Plutonium			$9.740\ 10^{3}$	36.9	226.3
Minor actinides:					
^{237}Np	$2.14\ 10^{6}$ y	$1.1\ 10^{-7}$	433.0	1.64	10.06
^{241}Am	432.6 y	$2.0\ 10^{-7}$	22.25	0.084	0.52
^{242m}Am	152 y	$1.9\ 10^{-7}$	0.731	$2.769\ 10^{-3}$	$16.98\ 10^{-3}$
^{243}Am	7380 y	$2.0\ 10^{-7}$	101.3	0.384	2.36
^{242}Cm	163 d	$1.3\ 10^{-8}$	0.131	$0.496\ 10^{-3}$	$3.04\ 10^{-3}$
^{243}Cm	28.5 y	$2.0\ 10^{-7}$	0.321	$1.22\ 10^{-3}$	$7.48\ 10^{-3}$
^{244}Cm	18.1 y	$1.6\ 10^{-7}$	24.0	0.091	0.56
Total Minor actinides			581.7	2.203	13.5
Some long-lived fission residues:					
^{99}Tc	$2.1\ 10^{5}$ y	$7.8\ 10^{-10}$	813.0	3.08	18.9
^{129}I	$1.57\ 10^{7}$ y	$1.1\ 10^{-7}$	169.5	0.64	3.92
^{135}Cs	$2.3\ 10^{6}$ y	$2.0\ 10^{-9}$	1307	4.95	30.35

a) 50 years dose commitment by ingestion in Sv/Bq, according to recent values given by the ICRP-68.
b) In one ton of Heavy Metal (tHM) of spent fuel discharged from a 900 MW(e) light water reactor (LWR), with a burn-up of 33 000 MW(th)*day/ton after 3 years cooling.[78]
c) The same, referred to the electricity production of 1 TWh (1 billion of kWh).
d) Referred to the annual output of a 1000 MW(e) LWR, at a load factor of 70 %.

Tab. 1. Production of the most important long-lived nuclei in spent fuels.

[78] In this chapter, wherever the unit W (watt) is used, the postscript "(e)" indicates "electrical", and the postscript "(th)" refers to "thermal".

Currently, public attention focuses mainly on the second of the above questions, dealing with present and future commercial wastes. The use of nuclear energy to produce either electricity for civilian use or nuclear materials (e.g. plutonium, tritium) for military purposes, generates indeed a certain amount of nuclear wastes. In most cases, the major waste related risk comes from their radiotoxicity, following the incorporation in the human body through food pathways or by external irradiation. The radiotoxicity is directly proportional to the activity of the various radionuclides held in these wastes, and can therefore last for very long periods of time, far beyond human societies' life-span, due to the possible presence of long-lived radionuclides. Almost all of these are artificially generated inside fission reactors, as a result of neutron irradiation.

Among them, actinides are largely responsible for the long term radiotoxicity component. They are produced through successive neutron capture (possibly followed by β-emission) and some (n,2n) reactions, starting from ^{235}U and ^{238}U present in the fresh enriched uranium fuel. Beside uranium isotopes, most of these actinides are trans-uranium nuclei. They include plutonium, the most abundant one usable as a nuclear fuel, and minor actinides (neptunium, americium and curium), considered as wastes. Most of these actinides are characterised by a long lifetime and, many being α-emitters, a high radiation dose factor (or "dose coefficient"; see table 1 in Chapter 1, by Hill and Pease). Some other long-lived radionuclides are also found among the fission and activation residues. They show much lower radiation dose factors, but their mobility in the environment being higher than that of actinides (which is usually the case), they must also be taken into account in any long-term management scheme.

Table 1 shows their corresponding inventory inside the spent fuel of a pressurised light water reactor (PWR). Approximately 40 kg (assuming that all ^{241}Pu has decayed to ^{241}Am) of minor actinides and 200 kg of plutonium are produced each year by such a 1000 MW(e) reactor. A cumulative production of tens of tons of long lived nuclear wastes is expected in the OECD (Organisation for Economic Co-operation and Development) countries by the year 2010, where one expects that approximately 180 000 tons of spent fuels will have been unloaded from nuclear reactors.[79] The contributions to the long-term potential risk by ingestion for each of the actinide elements present inside spent fuel are also shown in figure 1. As expected, plutonium isotopes are responsible for almost all the radiotoxicity, at least up to 100 000 years, when neptunium-237 becomes an important contributor through its descendants.[80]

---

[79] *Nuclear Energy Data*, report by the Nuclear Energy Agency (NEA) of the Organisation for Economic Co-operation and Development (OECD), Paris, 1993.

[80] Editors' note: the dose values indicated in figure 1 are purely hypothetical, being derived from a model in which the entire nuclear waste output of a reactor is assumed to be ingested by a human population.

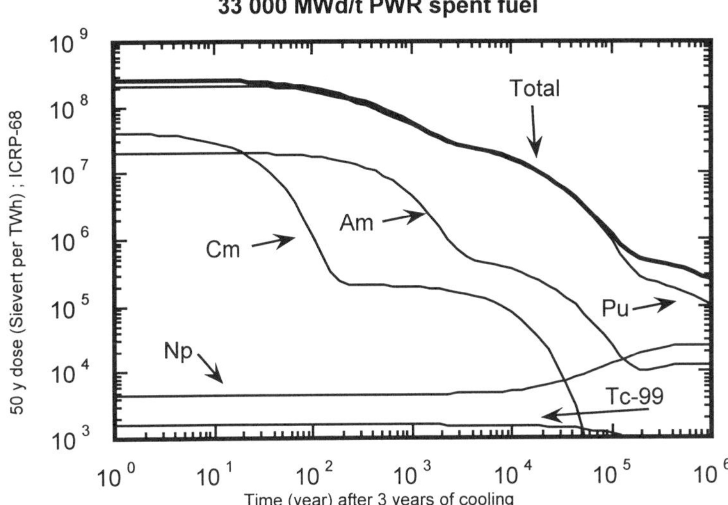

Fig. 1. Potential risk by ingestion for 33 000 MWd/t PWR spent fuel.

There are essentially two options to deal with the spent fuel unloaded from commercial nuclear reactors, after they have been stored in water-pools to let the residual heat decay for 5 to 10 years.

The first option is to reprocess the spent fuel, in order to separate uranium and plutonium. During this process, the remaining constituents, fission products and minor actinides, are left together as nitrate solutions and represent the high level and long-lived wastes. On the other hand, medium level waste (called B waste in the French classification), containing some long-lived nuclei (e.g. plutonium, I-129), is also produced (claddings, fuel assemblies, metallic structures, decontamination sludge).

The recycling of uranium in the reactors themselves, at the cost of a certain enrichment increase that is necessary to compensate for the negative effect of even isotopes (U-234, U-236) on the neutron economy, is not presently being considered on a large scale. This is not the case for separated plutonium, which is partially recycled in the so-called MOX (Mixed Oxide, uranium - plutonium) fuels loading some reactors. Recycling is done in France, where twenty 900 MW(e) PWR have presently 30 % of their cores loaded with MOX fuels, and also in Belgium, Germany and Switzerland; it is foreseen in Japan. Nevertheless, MOX fuel fabrication capacities, licensing situations and also the lack of economic incentives, probably explain the fact that more than 150 tons of civilian separated plutonium are stored, exceeding the needs for recycling (see table 2 in Chapter 13, by Goldschmidt).

High level wastes are vitrified in borosilicate glass, ready for final geological disposal after a long heat decay period of the order of 50 years. The so-called B wastes cannot be stored at the surface and are also intended for deep geological disposal, after being conditioned in various forms (bitumen, cement). The reprocessing option is commercially implemented in a few countries only, like France, UK, and Japan, and applies to about 20 % of the world-wide unloaded spent fuels (for the OECD countries).[81]

In the second option, interim storage takes place, usually near reactors, for a period of time which can extend to 50-100 years in dry storage casks. Thereafter, spent fuels are intended to be disposed of directly in a deep underground storage site (or eventually reprocessed). This option is adopted or considered in a number of countries like Sweden, the United States and Canada, whereas other countries are considering both options (e.g. Germany). The present spent fuel strategies in different countries are presented in table 2.

There is a consensus among the national and international nuclear agencies[82], that geological disposal is the best long-term solution either for vitrified packages and α-contaminated wastes (B wastes), or for spent fuels, and that it gives large margins of safety at any time in the future, provided that a suitable site has been properly selected. Besides the recent decision to store military high level wastes in a salt formation in New-Mexico (US)[83], there is nowhere in the world another such operating site. But important studies are carried out in underground laboratories (e.g. in Canada, Sweden, Switzerland, Belgium) to select and qualify favourable geological sites in various types of rock (granite, clay, salt, tuff in the case of the US).

---

[81] *Nuclear Energy data, op. cit.*

[82] That is: the EC (European Commission), the IAEA (International Atomic Energy Agency) and the NEA (Nuclear Energy Agency of the OECD).

[83] The so-called WIPP site.

Country	Nuclear share in electric production (%) a)	Installed capacity in GW(e) a)	Spent fuel management policy	Unloaded spent fuels (tons) b)	Unloaded spent fuels (tons) c)	Reprocessed spent fuels (tons) d)
France	76.4	58.5	R-D, (E)	1150	11770	2518
Belgium	55.8	5.5	R, E	120	1400	403
Sweden	51.1	10.0	ST-DIR	250	*≈3240*	
Switzerland	36.8	3.0	R, E	85	*≈1300*	285
Spain	35.0	7.1	E	168	1775	
Finland	29.5	2.3	E	70	*≈975*	
Germany	29.3	22.7	R, E	470	6315	2706
Japan	27.2	38.9	R, R-D	981	*≈8600*	1082
UK	25.8	11.7	R-D, E?	826		
US	22.0	98.8	ST-DIR	2200	*≈28600*	
Canada	19.1	15.8	ST-DIR	1479		
Netherlands	4.9	0.5	R, E	15	150	162

a) Situation at 1 January, 1995 (source: IAEA).
b) In 1992 (source: Nuclear Energy Agency of the OECD).
c) Spent fuel oxides cumulative production up to 1995 (source: EC); data in *italics* have been estimated from the accumulated nuclear electricity production (source: CEA, 1992) and taking an average value of 4.2 tons of discharged spent fuel per TWh(e) (case for Germany).
d) At the La Hague reprocessing plant, spent fuel oxides cumulative production up to 1 March 1995 (source: COGEMA). R-D: reprocessing in the same country; R: reprocessing outside the country; E: interim storage; ST-DIR: direct disposal of un-reprocessed spent fuel.

Tab. 2. Some useful data related to the back-end strategies in OECD countries.

2. Some Issues Related to Nuclear Wastes

This brief review of the present status of nuclear waste management points to some important issues to be addressed; some technical others political. In our opinion, the most important issues are the following ones:

- the management of spent fuels and of separated plutonium;
- the possibility of going a step further than the present reprocessing option, namely to the so-called partitioning-transmutation strategy;

- the relation between waste management and long-term nuclear development and the possible use of thorium-uranium fuel cycles instead of the present uranium-plutonium cycle.

3. The Management of Spent Fuels and of Separated Plutonium

The way nuclear wastes are dealt with in the back end of the fuel cycle depends on the strategy adopted with respect to spent fuel management, and more specifically to the fate of plutonium. As mentioned in §1, any long-term management scheme has to address the following plutonium issues :

1 - spent fuel radiotoxicity, which is almost completely determined by its plutonium content (see figure 1).

2 - the potential, for any type of plutonium produced inside commercial reactors, to become a weapon material and therefore needing to be subject to strong safeguards requirements.

3 - the potential for plutonium to be used as a fuel in thermal or fast reactors, because of the presence of two highly fissile isotopes (Pu-239 and Pu-241) and the large amount produced (12 tons/year for the French nuclear programme).

In principle plutonium is considered as an asset (reason 3) and not as a burden (reasons 1 and 2) in the nuclear industry, at least in countries strongly committed to reprocessing. From a physical point of view, a fast breeder is the most efficient way to use plutonium, due to a neutron economy balance and a fission-to-capture ratio more favourable than in thermal reactors. But, fast breeder reactors do not exist on an industrial scale and are not expected before the second half of the next century, for economical reasons related to the enriched uranium market and to the stand still of nuclear programs in many industrialised countries. The present context leads in fact to the idea that plutonium is a burden, at least for the time being, and that it must be neutralised by mixing it with highly radioactive materials and disposed of or possibly be burnt.

If there is no reprocessing, plutonium is considered safely sequestered, at a concentration of the order of 1 %, in the highly radioactive environment of the spent fuels.

On the other hand, if plutonium is chemically separated, the only possibility left is its recycling in the present thermal reactors, in the so-called MOX or MIX fuels (both being mixed plutonium - uranium oxide fuels: see table 3 for the explanation of the difference between MOX and MIX). Different ways of plutonium recycling are presented in table 3 together with the open cycle carried out with the new so-called N4 reactor type adopted in France. All these recycling scenarios imply MOX fuel fabrication plant and refitting of present light water reactors (LWR) in order to satisfy safety and operational requirements. When more than one recycling is contemplated, substantial modifications have to be made at the reprocessing plant to take into account higher plutonium concentrations as well as increased radiation and thermal constraints (e.g. increase of americium, curium and Pu-238).

The data given in table 3 call for the following comments:

a) before any recycling, one can first tune the reactors in such a way as to reduce the amount of plutonium produced inside the spent fuel. Burn-up increase, which represents a strong economical incentive, has such an effect but at the expense of increasing the minor actinides. On the other hand, increasing the moderating ratio also has a strong effect, not only on plutonium reduction but also on the minor actinides. Although it increases the size of the reactor at the same power level, it brings economical gain on the U-235 enrichment (see table 3, column 2). The design of new reactors should take these possibilities into account.

b) in the first recycling scheme, either in a PWR or an FBR (Fast Breeder Reactor) - see table 3 (columns 3 & 4) - plutonium is mixed at a high concentration level with depleted or natural uranium in the so-called MOX fuels. In such an heterogeneous scheme, plutonium is recycled in some specialised reactors. For safety reasons, only 30 % of the core of such a reactor is loaded with MOX fuels. Unfortunately, MOX recycling leads to an important build up of even plutonium isotopes and of minor actinides, as a consequence of the thermal spectrum (see table 3, column 3). But, as far as plutonium is concerned, there is an important net burning (essentially Pu-239, the fissile isotope) during the first recycling in the MOX fuel assemblies, but not in the entire core, due to plutonium production inside the standard enriched uranium fuel assemblies. Since a first recycling carries some technical and economic penalties, a second recycling cannot really be considered, and one is left with MOX spent fuels. In any case, multi-recycling, which would lead to a plutonium inventory stabilisation, is not conceivable for safety reasons, because plutonium enrichment would have to be increased at each new cycle, leading to a positive void coefficient.

The gain with such an heterogeneous scheme is only in the short term, because the plutonium concentrated in the highly radioactive MOX spent fuels (around 5 to 7 % as compared to 1 % in so-called UOX, enriched uranium oxide, fuels) is isotopically degraded. After a period of interim storage, lasting 40-50 years, the spent MOX fuels could be reprocessed in order to use the degraded plutonium in fast reactors or to enable disposal deep underground. Nobody can presently predict which strategy will then be adopted but one knows that long-term radiological and thermal impacts (important considerations for the deep disposal site) of spent MOX fuels are much more important than those of UOX fuels. In our opinion, it appears that, in the long-term, MOX strategy leads to a deadlock, or to reliance on very uncertain predictions concerning the use of such fuels in fast breeder reactors.

Scenario:	**Open Cycle**		**Heterogeneous Recycling**		**Homogeneous Recycling**
Column number	1	2	3	4	5
Reactor type	PWR N4	PWR N4	PWR MOX	FR CAPRA	PWR MOX
Moderating ratio	2	3	2	/	2
Burn-up (MWd/t)	55 000	55 000	55 000	140 000	55 000
Cycle number	1	1	1	Equilibrium	Equilibrium
Loaded fuel:					
Type of fuel a)	UOX	UOX	MOX	MOX	MIX
Plutonium content (%)	0	0	10	54	2.0
^{235}U content (%)	4.5	3.8	0.25	0.19	3.8
Mass balance, variation at 5 years after unloading in kg per TWh(e):					
Plutonium	+29	+21	- 66	- 87	+0
Neptunium	+2.1	+1.4	+0.2	+0.3	+1.6
Americium	+1.4	+1.4	+14	+16	+4.5
Minor actinides total	+3.8	+2.9	+17	+18 b)	+8.3
Introduction of such reactors in a park of N4 reactors:					
Proportion in % in the park c)	100	100		24	100
Total MA production in kg/TWh(e) d)	+3.8	+2.9		+6.9	+8.2
Total Pu net production in kg/TWh(e)	+29	+21		+0	+0
Total Pu inventory in the fuel cycle (ton) e)				310	200

a) UOX: enriched Uranium OXide; MOX: Mixed OXide plutonium + depleted or natural uranium; MIX: MIxed oXide plutonium + slightly enriched uranium.
b) Mainly coming from Am-241 build up from β-decay of the large Pu-241 inventory.
c) In all cases, the standard reactors of the park are of the N4 - UOX type (see first column).
d) MA: minor actinides (neptunium, americium and curium).
e) One assumes that the total mixed reactor park has a power of 60 GW(e), produces 400 TWh(e) per year and that the cooling time before reprocessing is 3 and 5 years for PWR and FBR fuels respectively.

Tab. 3. Some Plutonium Recycling Scenarios.
Source: Commissariat à l'Energie Atomique (CEA).

c) In fact, there are two ways to multi-recycle plutonium, either in fast reactors tuned to incinerate plutonium (such as the so-called CAPRA concept[84], developed by the CEA: see table 3, column 4), or in thermal reactors in an homogeneous way, that is to say

[84] The abbreviation CAPRA, the French project designed to burn high levels of plutonium, stands for "*Consommation Accrue de Plutonium dans les réacteurs à neutrons RApides*".

where each reactor takes care of its own plutonium (see table 3, column 5). In this case, all reactor cores are loaded with MIX fuels using slightly enriched uranium and a low plutonium concentration. In these scenarios, after a period of about 60 years of recycling, plutonium inventories are stabilised at levels of 310 and 200 tons respectively. These scenarios imply a long-term commitment to nuclear energy and require very strong safeguards because large plutonium quantities are separated each year when equilibrium is reached. Contrary to widely accepted views, homogeneous multi-recycling in thermal reactors has better features in terms of actinide inventory than is the case in fast reactors.

d) The same time constants exist if one wants to reduce the plutonium inventories; in that case one has to burn plutonium without any uranium in the fuel in order to reach the maximum burning limit of 46 kg per TWh(th). This would probably require specialised burners as discussed in §4.

In conclusion, it is very difficult in the present nuclear energy context to stabilise plutonium inventories in a commercial pool of reactors and, if so, it would lead to strong safeguards concerns. It seems that for the time being mono-recycling or no reprocessing at all are acceptable, knowing that in the long-term many uncertainties (related to fast reactor programs, or to the deep disposal strategy) remain.

4. Transmutation as a Complementary Option to Geological Disposal?

There is a general consensus among the national and international organisations in charge of nuclear energy development, that final geological disposal of highly active and long-lived nuclear wastes represents the reference management option, and that it can be safely implemented.[85] However, since the 1990s, there has been a renewed interest in the so called P-T strategy (Partitioning-Transmutation), entailing chemical separations of elements having some long-lived isotopes (> 30 years), in order to transmute them into shorter lifetimes or stable nuclides. In Japan, a very ambitious program, called OMEGA, has been proposed, to go far beyond the standard reprocessing option, including P-T. As a consequence of strong oppositions to the construction of underground laboratories aimed at qualifying a disposal site, a law has been adopted in the French Parliament (December 30, 1991) providing for a research program on long-lived nuclear waste management including P-T.

Indeed, the P-T option could be applied to some long-lived radionuclides (actinides, fission products) in order to address some concerns related to long-term effects of deep geological disposal on man and his environment, and to public acceptance. One can single out the main arguments which are usually given in favour of the transmutation concept:

[85] These international organisations are notably the EC, the IAEA and the NEA (OECD).

- weak reliability of long-term predictions;
- safety improvement of geologic disposal;
- reducing the thermal impact of wastes on geological repository;
- shortening the period of concern about the waste repository;
- ALARA principle[86], and responsibility for future generations of mankind;
- utilisation of useful material for reactor operation;
- enhancement of public acceptability of nuclear power.

With the present technologies available to achieve partitioning and transmutation, one will probably not be able to avoid a certain kind of geological disposal, making P-T appear more as a complementary option than an alternative. However, some very innovative concepts have been proposed for which long-term geological storage could be avoided (for example the Energy Amplifier proposed by Carlo Rubbia: see Chapter 16, by van der Zwaan).

Transmutation can take place in thermal or fast reactors; it usually needs, as for plutonium reduction, long irradiation times, which means in practice many recycles (at least for solid fuels). With present technologies and using fast reactors, radiotoxicity can be theoretically reduced by a factor ranging between 10 and 100, if minor actinides are continuously recycled together with plutonium.[87] The destruction of long-lived fission products (see table 1) is more difficult, because excess neutrons must be available in the epithermal neutron energy region. In practice, Tc-99 and I-129 are the only long-lived fission products that need to be considered for destruction in reactors.

Because destruction and production of radionuclides are normally in approximate balance, recycling leads to an equilibrium inventory. Low equilibrium inventories are theoretically achieved with dedicated fuels, that is to say without plutonium or uranium if one wants to destroy minor actinides, or without uranium in case plutonium burning is the goal. Loading normal reactors with such exotic fuels may need to be limited for safety reasons; recent proposals such as the CAPRA concept applied to fast reactors or using accelerator-driven subcritical reactors tend to avoid such difficulties and will probably play a very important role in case efficient transmutation is implemented on an industrial basis.

There is another issue related to secondary wastes. In fact, transmutation normally implies chemical partitioning and multiple recycling, during which other types of wastes containing small fractions of actinides are created. Unless high decontamination factors are achieved (one order of magnitude at least compared to present practices), the long-term impact of these wastes might overcome the benefit of transmutation, which affects only the high level waste packages.

Finally, any transmutation strategy has to be consistent with the way other types of long-lived wastes are managed. For example, one has to compare the real risk avoided through transmutation with the long term risk due to uranium tailings (Th-230, Ra-226), taking into account their long-term management schemes.

---

[86] ALARA stands for "As Low As Reasonably Achievable".

[87] M. Salvatores, hearings at the *Commission Nationale d'Evaluation*, November 1995.

5. Waste Management in Relation to Long-term Nuclear Development and the Use of Thorium

The important issue concerning the long-term impact of nuclear wastes might be solved in theory by destroying these wastes in order to reduce the time during which institutional monitoring is required. The difficulty with such a strategy is that it implies a very long-term commitment to nuclear energy (at least 100 years) in order to have the funding, the industrial background and the motivation to go on with sophisticated technical systems like those needed for large scale transmutation, such as with fast reactors, accelerator-driven systems (ADS), or new reprocessing methods like pyro-metallurgy, and so on. Nuclear energy is not by any means at this stage today, when its share seems diminishing in most OECD countries, whereas it increases in Asian countries for which transmutation is not a first rank priority. In this context, waste management can only be modest, long-term interim storage being the key word for the next few decades. Meanwhile, deep geological disposal (for all types of foreseen wastes, including spent fuels), together with more advanced waste conditioning and transmutation, can be carried out at a R&D level, in order to prepare for the future.

In this respect, some analysts foresee that, in the second half of the next century, nuclear energy might develop on a new basis as a result of strong concerns related to resources (fossil, U-235) as well as to environmental protection (e.g. CO_2 emission).

In that context, the use of thorium for energy production could be an important option. Thorium is three times more abundant than uranium and is a fertile nucleus from which the new fissile material U-233 is bred. During the irradiation of a thorium based fuel, the spectrum of actinides produced during neutron irradiation is completely different from that produced from an uranium based fuel. In particular, there is virtually no plutonium produced and the lifetimes of the various thorium, protactinium and uranium isotopes are such that the radiotoxicity during the first 10 000 years is smaller than that of the uranium based fuel by more than one order of magnitude. On the other hand, management of high level wastes either for interim storage or for final disposal becomes easier, because thermal impacts are much smaller than those of the normal uranium-plutonium fuel cycle. Long term effects are also much smaller in the front end mining tailings, due to the short half life of Ra-228 (5.7 years) as compared to the half life of Ra-226 (1600 years). The impact of radon is also less with thorium mining than with uranium. For all these reasons, since the end of the 1980s there has been a renewed interest for thorium, possibly used with accelerator driven systems.

Conclusion

One important demand within society is avoidance of irreversible commitments in those matters, which are perceived as very uncertain. In this respect it is fair to say that interim storage of spent fuels leaves the way open to a retrievable solution in the sense that it allows one in the future to chemically separate valuable elements, such as plutonium or some noble metals, or specific ones (actinides, fission products) to be transmuted. Retrievability in the case of deep geological disposal is also considered, at least for a certain period of time, when the disposal site would be monitored before closing in a definite way. All this implies that, to a certain extent, future generations have to share the nuclear waste management with the present generation. Recognising that there will never be a completely safe solution to long-term waste management, what remains to be done today is to take the best options in a retrievable way in order to minimise the burden on future generations. In this perspective, interim storage cannot be a strategy for a long and indefinite period of time, but rather for a limited period (e.g. 50 years), possibly renewable if society so decides. Long-term waste management cannot rely only on technical actions but has to be responsibly taken in charge by society itself, now and in the future.

Chapter 9 The Storage of Nuclear Wastes

by
John L. Finney*

Introduction

The previous chapter by Schapira has provided a broad overview of the nature of nuclear wastes and of some of the strategies, both current and speculative, that may need to be adopted for managing them. Here, we shall focus predominantly on storage strategies and the issues that they entail, concluding with a brief discussion on the role that transmutation techniques might play.

In 1976, the UK Royal Commission on Environmental Pollution recommended that there should be no commitment to a large programme of nuclear power "until it has been demonstrated beyond reasonable doubt that a method exists to ensure the safe containment of long-lived highly radioactive waste for the indefinite future".[88] Similar statements have been made with respect to other nuclear power programmes, with Sweden passing a law in the mid-1970s stipulating that the country's nuclear power programme could not continue unless it could be shown that high level waste could be disposed of safely. Considering the potential effects of a failure to deal adequately with such wastes, such a demonstration seems entirely reasonable before accepting nuclear power as a long-term option.

During the relatively short history of nuclear power, a legacy of waste has been built up that must be dealt with effectively, irrespective of whether or not we embark on an expanded programme of nuclear power. A solution must be found for the waste that already exists. However, a solution that may be appropriate for the waste inventory created by presently operating stations *may* not be appropriate for a continuing programme, for example for volume reasons. Moreover, a continuing programme might use a different fuel cycle that produces waste which is less problematical to deal with.

This chapter tries to assess the current waste situation, the options that have been proposed for dealing with it, and some of the scientific problems of these options. Solving the problem is very much an interdisciplinary task, involving (often complex) physics, chemistry, geology, and materials science. This chapter cannot therefore hope to be exhaustive. I have chosen to focus on what my study of the issues suggests are the critical problems. Moreover, although international experience is discussed, the emphasis is primarily in the UK context. This is appropriate in that the UK problem is at least as technically difficult as any faced by any other country, although it should be borne in

---

* John Finney is Professor of Physics at the Department of Physics and Astronomy, University College, London.

[88] UK Royal Commission on Environmental Pollution, 6th Report 1976, Cm 6618.

mind that some of the problems faced in the UK may not be as severe elsewhere, e.g. in simpler geological situations.

1. The Nature of Radioactive Waste from Nuclear Power Reactors

A nuclear power programme generates radioactive waste primarily from:

- materials and equipment that have become contaminated during power station operation, and the manufacture of nuclear fuel;
- decommissioning of reactors at the end of their lifetimes (~ a few tens of years);
- spent fuel, or - if a reprocessing system is in operation to separate uranium and plutonium, e.g. for possible reuse - waste arising from reprocessing.

Mainstream reactors burn UO_2 fuel, the enrichment of the fissile ^{235}U being typically 3-4 %. Fission of ^{235}U results in isotopes such as ^{90}Sr, ^{137}Cs and ^{129}I, while the "matrix" ^{238}U captures neutrons to accumulate an inventory of transuranic isotopes, prevalent in which are a number of plutonium isotopes. We note here that some of the major problems of waste are generated not by the fuel itself, the fissile ^{235}U, but by the matrix material, usually ^{238}U. Other matrices may be useable, with implications for possibly easing the waste situation. Reprocessing of the spent fuel, when undertaken, separates the potentially reusable uranium and plutonium from the residual waste fission and actinide products.

Classifications of waste do not appear to be identical internationally. Though characterisations might usefully relate to the kind of activity (α, β, γ), longevity, and toxicity, emphasis is also placed on heat production. The UK classification of very low level waste (VLLW), low level waste (LLW), intermediate level waste (ILW) and high level waste (HLW) is set out in table 1.[89]

Category	**β and γ activity (Bq kg^{-1})**	**α activity (Bq kg^{-1})**
VLLW	< 400	< 400
LLW	400 - 1.2 x 10^7	400 - 4 x 10^6
ILW	> 1.2 x 10^7	> 4 x 10^6
HLW	No specification	No specification: so active as to generate significant heat

Tab. 1. UK classification of radioactive nuclear waste activity.
Source: UK Department of the Environment, 1995, op. cit.

[89] UK Department of the Environment, 1995, *Review of Radioactive Waste Management Policy*, Final Conclusions, Cm2919.

The detailed radionuclide content of the waste depends on the kind of reactor used, the reactor operating conditions, and on any post-treatment (e.g. reprocessing). For example, table 2 lists the radionuclide content of UK intermediate level waste and more active low-level waste expected in 2030, together with the half-lives and activity.[90]

Although the waste constitution will vary with a number of factors, the list of table 2 is representative enough to show the range of half-lives (from a few tens to 10^{10} years) and the kind of relative activities expected of the various nuclides. For example, though ^{232}Th may have a 10^{10} year lifetime, the activity level is relatively low. However, combinations of high activity and long half-life are found with some nuclides, such as ^{99}Tc (2 x 10^5 years and 170 TBq) and ^{239}Pu (2 x 10^4 years and 104 TBq). Furthermore, some of the most active waste may have relatively short lifetimes, examples being ^{90}Sr (29 years and 3 x 10^5 TBq) and ^{137}Cs (30 years and 9 x 10^5 TBq). Effective strategies for dealing with this waste therefore need to take account of this range of activity and half-life. For example, although the half-life of ^{137}Cs is relatively short, its activity in the expected UK waste content is such that its activity level will take 200 years to decay to that of the much-longer lived ^{239}Pu. We need to take account of both activity and decay rate.

Radionuclide	**Half-life (years)**	**Activity in waste (TBq)**	**Radionuclide**	**Half-life (years)**	**Activity in waste (TBq)**
Tritium (^{3}H)	12	1.0 x 10^5	Radium-226 (^{226}Ra)	1.6 x 10^3	8.6
Carbon-14 (^{14}C)	5.7 x 10^3	2.3 x 10^3	Thorium-230 (^{230}Th)	1.7 x 10^4	0.081
Chlorine-36 (^{36}Cl)	3.0 x 10^5	5.2	Thorium-232 (^{232}Th)	1.4 x10^{10}	0.058
Cobalt-60 (^{60}Co)	5.3	5.7 x 10^5	Proactinium-231 (^{231}Pa)	3.3 x 10^4	0.019
Nickel-59 (^{59}Ni)	7.5 x 10^4	1.2 x 10^4	Uranium-234 (^{234}U)	2.4 x 10^5	62
Selenium-79 (^{79}Se)	6.5 x 10^4	1.6	Uranium-235 (^{235}U)	7.0 x 10^8	1.5
Strontium-90 (^{90}Sr)	29	3.2 x 10^5	Uranium-238 (^{238}U)	4.5 x 10^9	34
Zirconium-93 (^{93}Zr)	1.5 x 10^6	3.4 x 10^2	Neptunium-237 (^{237}Np)	2.1 x 10^6	33
Niobium-94 (^{94}Nb)	2.0 x 10^4	1.4 x 10^3	Plutonium-239 (^{239}Pu)	2.4 x 10^4	1.1 x 10^4
Technetium-99 (^{99}Tc)	2.1 x 10^5	1.7 x 10^2	Plutonium-240 (^{240}Pu)	6.5 x 10^3	1.2 x 10^4
Tin-126 (^{126}Sn)	1.0 x 10^5	4.0	Plutonium-241 (^{241}Pu)	14	3.0 x 10^5
Iodine-129 (^{129}I)	1.6 x 10^7	0.92	Plutonium-242 (^{242}Pu)	3.8 x 10^5	14
Cesium-135 (^{135}Cs)	2.3 x 10^6	8.8	Americium-241 (^{241}Am)	4.3 x 10^2	3.5 x 10^4
Cesium-137 (^{137}Cs)	30	9.3 x 10^5	Americium-243 (^{243}Am)	7.4 x 10^3	31

Note 1: volumes of waste are 3 x 10^5 m^3 ILW and 7 x 10^4 m^3 LLW.
Note 2: Measures of radiotoxicity of many of the above radionuclides ("dose coefficients") are listed in table 1 of chapter 1, by Hill and Pease.

Tab. 2. Radionuclide content of UK ILW and more active LLW in the year 2030.
Source: UK Nirex Ltd, 1993, op. cit.

Current practice accepts that the potential doses from VLLW are so low compared to natural exposure that it can be disposed of as normal domestic waste without treatment.

90 UK Nirex Ltd, 1993, *Nirex deep waste repository project*, Scientific update 1993, Nirex report no. 525.

LLW is similarly argued to require no shielding, and again generally accepted practice is that it can be compacted, placed in containers, and buried in shallow sites.

The main problems are how to deal effectively with ILW (often material that has been in contact with active materials, results from reprocessing, and arises during reactor decommissioning) and HLW. The activity levels of the former require shielding and special handling. The latter is so active that significant excess heat is generated. In the UK and some other countries, this is vitrified in borosilicate glass and stored above ground for tens of years to cool before final disposal is decided upon. The volume of HLW is relatively low; although in the UK context it makes up about 1 % of the total radioactive waste volume, it contains nearly 75 % of the radioactivity in the irradiated fuel.[91]

Longevity of activity is not considered in the above classification. HLW by its nature contains long half-life materials, while ILW is a mixture of long and short (where the short/long distinction tends to be made at a half-life of 30 years - with a "short-lived" isotope such as ^{137}Cs decaying to 0.1 % of its initial activity after 300 years). Thus, it is often considered useful to divide the ILW category into short-lived and long-lived radionuclides. Disposal strategies of most countries distinguish ILW in this way, and lump the long-lived ILW and HLW together; exceptions include Germany and the UK.[92] With such a separation, it has been proposed that the short-lived ILW could be stored (e.g. surface, or recoverably underground) for a few hundred years to allow the radioactivity to decay, so that it can ultimately be disposed of as LLW.[93]

We should note also that some of this "waste" - the spent fuel - is still an energy resource: in the context of thermal reactor technology a spent fuel rod still contains 60 to 75 % of its energy content. Reprocessing separates off the plutonium and uranium for possible recycling (e.g. as "Mixed Oxide" $(U,Pu)O_2$, MOX, or for use in fast reactors), though non-proliferation arguments have persuaded some countries (e.g. the US) not to go down this route. In such cases, it is proposed to dispose of unprocessed spent fuel as waste.

2. Existing Strategies

2.1 The regulatory framework

Internationally applicable (though not binding) standards for radioactive waste have been formalised through the International Commission on Radiological Protection

---

[91] UK Department of the Environment, 1997, *Digest of environmental statistics*, No. 18.

[92] The Royal Society, 1994, *Disposal of radioactive wastes in deep repositories*, Report of a Royal Society Study Group, pp. 69-70, and UK Nirex Ltd, 1993, *Going underground - how other countries dispose of their radioactive waste.*

[93] UK Parliamentary Office of Science and Technology, 1997, *Radioactive waste - where next?*, p.5.

(ICRP). The latest (1990) revision - itself currently being revised - states some general principles covering radiation protection *per se*.[94] These include:

- *justification*: an adopted radiation practice should produce a positive net benefit;
- *optimisation*: exposures should be kept as low as reasonably achievable (ALARA);
- *limitation*: doses to individuals should not exceed recommended limits.

Furthermore, the 1992 "Earth Summit" agreed that principles of sustainable development should apply. In particular with respect to waste management, (a) future generations should not be left with burdens and risks created by current generations; (b) sound science, and (c) the precautionary principle should apply.[95] Both the ICRP principles and the sustainable development requirements have subsequently been incorporated by the International Atomic Energy Agency (IAEA) into nine guiding principles on the management of radioactive waste.[96]

In the UK, the recommended criteria to be met by a post-closure performance assessment (PCPA) of a potential disposal site include a target of 1 in 10^6 per year as the maximum additional risk to an individual of contracting a fatal cancer or hereditary defect, with a risk greater than 1 in 10^5 being considered as unacceptable.[97,98] These risk factors compare with an average equivalent risk from exposure to natural radiation of 1 in 10^4, and a maximum level of exposure for members of the public from a nuclear installation of 1 in 10^5.[99] The ALARA (as low as reasonably achievable) and ALARP (as low as reasonably practical) principles also apply.

2.2 Low and very low level waste

As mentioned above, the activity of this (generally dilute) material, together with the timescales involved, are such that potential radiation doses are believed to be insignificant. Good engineering practice should be sufficient to ensure whatever degree of temporary environmental isolation is needed, and eventual dispersion is high enough to raise no problems with the regulatory guidelines. I will not discuss further VLLW or LLW.

---

94 ICRP, 1990, *Recommendations of the International Commission on Radiological Protection*, ICRP publication 60, Ann. ICRP, 21, no. 1-3.

95 OECD Nuclear Energy Agency, 1995, *The environmental and ethical basis of geological disposal of long-lived radioactive wastes*.

96 IAEA, 1996, *IAEA Safety Fundamentals: the principles of radioactive waste management*, Safety series no. III-F.

97 UK National Radiological Protection Board, 1991, *Living with radiation*; UK National Radiological Protection Board, 1992, *Statement on radiological protection objectives for land-based disposal of solid radioactive wastes*, Documents of the NRPB, Vol. 3, no. 3.

98 UK Parliamentary Office of Science and Technology, 1997, *op. cit.*, pp. 7-10, 19-23.

99 UK National Radiological Protection Board, 1991, *op. cit.*, and UK National Radiological Protection Board, 1992, *op. cit.*

2.3 Short-lived ILW

Separating off the short-lived from the long-lived ILW allows advantage to be taken of the shorter timescales in any disposal strategy. The main problem short-lived isotopes are ^{137}Cs and ^{90}Sr. Activity reduction factors of 1000 are achieved over 300 years for the longest-lived component with a 30-year half-life. At these timescales, it would seem reasonable to expect that adequate monitoring - and necessary periodic repackaging - could take place if storage were on the surface, or containment could be designed to last for such lengths of time. An advantage of medium-term retrievable storage is that options are not closed off. However, some views (e.g. from the OECD/NEA, Organisation for Economic Co-operation and Development / Nuclear Energy Agency[100]) consider this to be outweighed by the natural tendency of society to become accustomed in time to the presence of storage facilities, with the resulting complacency increasing the risk of accidents. While it is not a purpose of this chapter to list accidents that have already occurred to material in storage (e.g. Hanford in the US, Dounreay in the UK), I consider this a (albeit highly relevant) regulation and control problem, and not an inherently scientific or engineering one. Swedish work has concluded that it is technically feasible to construct containers of sufficient integrity that the waste will remain contained for 500 years, by which time activity will have fallen by a factor of 10^5.[101] Whether or not this results in an acceptable activity level as the material begins to disperse in the environment depends on factors such as the expected effectiveness of the remaining barriers to dispersal, and the original degree of activity of the amount stored - after all, 0.001 % of a very large number can still be significant, and the UK's "activity burden" in 2030 of ^{137}Cs and ^{90}Sr is high - see table 2. These points relate to the problem of disposal of the long-lived ILW and HLW discussed below.

2.4 HLW and long-lived ILW

Here, the timescales for activity decay are much longer, and there is a significant heat burden in the initial years for HLW. General strategy at present is to store HLW for (nominally) 50-100 years to cool, and also to enable a final disposal procedure to be decided on, perhaps together with the long-lived ILW as proposed in several countries.[102] Again, provided surveillance is effective - again not a scientific issue - this strategy would seem reasonable. The scientific issues include those related to the effectiveness of the containment, and the possible need for repackaging during the storage period. Immobilisation in borosilicate glass is widely practised, and considered technically acceptable[103] and other storage options include incorporation into a ceramic material. Although such procedures are generally thought to be adequate, we should recall that glass is a metastable material. Although structural measurements do show it does not

---

[100] OECD Nuclear Energy Agency, 1995, *op. cit.*
[101] The Royal Society, 1994, *op. cit.*, p. 40.
[102] See, for example, The Royal Society, 1994, *op. cit.*
[103] The Royal Society, 1998, *Management of Separated Plutonium*.

remain unaffected under radiation conditions[104], adequate monitoring during storage should be able to detect any problems of glass matrix deterioration before they become a problem.

Dealing with this category of material would appear to be the central waste problem that needs to be solved. Much work has been done on problems relating to this, and there is much literature - unfortunately some of it commercial in confidence - on which to assess the viability of proposed procedures. In addition to deep geological disposal - in which most effort seems to have been invested - there are other potential options which need to be considered. The rest of this chapter attempts to summarise the scientific issues relating to these options, which are, in addition to disposal, recycling in an appropriate power-generating facility, and transmutation.

3. Long Term Storage/Disposal

Effective disposal would be in a system that would prevent the significant entry into the biosphere of wastes at sufficient concentration to breach the set guidelines (see above). The half-lives of many of the isotopes involved mean these conditions must be maintained for over 10^6 years, a time scale that is difficult to relate to in human terms. Options that have been actively considered include surface storage, disposal on the seabed, burial under the seabed, and in deep (over several hundred metres) geological repositories in suitable rock formations. Sea dumping has been stopped by international agreement. This agreement comes due for review in 2019. This is only a short time ahead compared with the time scales characterising the basic disposal problem, and so seabed options should perhaps not be excluded. However, little work seems to have been done on these, and it would seem unlikely that disposal on the seabed itself would be considered acceptable, despite the ability of the sea to dilute. Under-sea bed disposal would seem to relate to that of geological disposal on land, so would appear to have common problems to be addressed.

Does such a site exist, in which very long-lived waste can be placed so that radioactive contaminants do not - over this very long time scale - return in excess of the set guidelines? Much has been made of studies of natural analogues, including the uranium deposits in Gabon that "went critical" some 1.7×10^9 years ago.[105] Here, despite there being local water circulation, there seems to have been only limited movement of radionuclides either during or after the reactor was operating, although there has been loss of alkali metals and some alkaline earths.[106] Bitumen and clay minerals in the surrounding rock apparently helped to retard their migration. This example and other less dramatic

[104] S. Susman *et al.*, 1990, "Structural changes in irreversibly densified fused silica - implications for the chemical resistance of high-level nuclear waste glasses", *Phys. Chem. Glasses*, vol. 31, p.144. See also A.C. Wright *et al.*, 1992, *J. Non-Cryst. Solids*, vol. 150, p.69.

[105] *Nature*, "The last natural nuclear fission reactor",1997, Vol. 387, p.22, and Uranium Information Centre Ltd (Australia), 1997, *Nuclear electricity - an Australian perspective*.

[106] R. Hagemann and E. Roth, 1978, "Relevance of the studies of the OKLO Natural Reactor to the storage of radioactive wastes", *Radiochimica Acta*, Vol. 25, pp. 241 - 247.

ones do demonstrate that geological disposal is in principle an option: there are sites that have been appropriate in the past, and there is no reason to suggest that other sites exist which will fulfill the requirements for a similar time into the future. The problem is to identify them now, with sufficient confidence.

Three main rock types have been considered as potential repository sites: crystalline rocks, clays, and salt. Each has its advantages with respect to the characteristics that are thought to be relevant, but each also has potential disadvantages. A major issue relates to groundwater flows (see below), so it might be thought that salt deposits would be particularly promising: they have after all been there a long time in the absence of water. Even here, however, there have been potential problems raised, including the possibility of containing pockets of corrosive brines, accidental flooding in the future, possible geological mobility of the formation, and the fact that it is an economic resource that might be mined in the future if records were lost.[107]

A great deal of work has been undertaken in relation to the identification of suitable repositories. We now consider some of the scientific issues relating to this identification, and some of the uncertainties involved. We (as has become customary) subdivide the disposal system into a number of different regions or zones, namely:

- the *near-field*: the wastes, their packaging, any surrounding, and the rest of the repository;
- The *geosphere*. This stretches from the repository's surrounding rocks, out to the more remote geological "far-field" region, up to and including the soils and other exposed material that contacts the biosphere. The geosphere includes the groundwaters within the rocks.
- The *biosphere* consists of seas and seabeds, soils, the atmosphere, and the organisms within them.

The problem of identifying a suitable repository is to convincingly demonstrate that the activity that may be released from the repository and transported to the surface environment over the relevant time scale will not exceed the agreed limits (see above). The relevant timescale for the wastes under consideration is at least 10^6 years. Natural analogue studies tell us such places are almost certain to exist. We must convince ourselves that our understanding of the various processes likely to occur in the disposal system over this time scale is sufficient to be adequately certain that our site identification is good.

The processes we need to consider are many and complex, and involve a range of science including physics, chemistry, geology, materials and environmental science. Characteristic of many of the problems is their multidisciplinarity, in that they cross traditional scientific boundaries. This is both a strength and a weakness. A strength lies in the range of different approaches that can be brought to complex problems, increasing the likelihood of imaginative approaches and solutions. A weakness - at least in the UK context - is that funding across disciplinary boundaries is often difficult.

[107] UK Parliamentary Office of Science and Technology, 1997, *op. cit.*, p. 77.

Some of the issues which are being addressed include diffusion and flow of groundwater; gas and two phase gas/liquid flow in fractured rocks; the possibilities of human intrusion, either directly into the repository or indirectly through e.g. drilling into the contaminated geosphere; natural disruptive events such as earthquakes which could lead to further rock fracture; and climatic changes which could lead to changed groundwater flow patterns. These are in addition to (what are probably easier to predict) the behaviour of engineered components and the integrity of the repository structure.

We cannot hope to identify a suitable repository by direct experiment. Modelling the processes involved is an obvious way forward. We might note that, as computational resources have improved consistently and extensively over the past decade, our ability to model increasingly complex systems, at both the molecular level and above, has increased dramatically. In the waste disposal context, however, this modelling is hugely complicated by the long time scales involved. It is a challenging enough problem to understand and model effectively the relevant processes when the characteristics of the system remain constant. But on top of that, we need to consider how the system's relevant characteristics may possibly change within that time. The uncertainties in what may happen over this timescale are such that we cannot hope to predict what will happen. We can only expect to aim to identify possibilities, and to attempt to quantify - or put bounds on - uncertainties.

A detailed discussion of the current state of modelling of processes relating to the viability of potential deep repositories is not appropriate in a chapter of this length. A well-considered appraisal of the state of affairs in the UK context in 1994 is given in reference.[108] I attempt below a summary of some of the issues and problems involved, and some of the implications for repository identification.

3.1 Process modelling

To develop and implement an effective model of a process - be it of groundwater flow in fractured rock, or of the atomistic behaviour of ions in aqueous solution (both in fact relevant to the waste disposal issue) - requirements including the following need to be satisfied.

- An understanding of the underlying processes. For example, in the ions in solution case, we need to be able to describe with good confidence the ways in which the components interact at the molecular level.
- An adequate characterisation of the system to be modelled. This includes in the groundwater flow case a specification of the flow channels, their geometries, and the characteristics governing the flow though them (e.g. fluid conductivity). What "adequate" means is not always easy to specify, though validation of a model gives some kind of justification of adequacy.
- A specification of the boundary conditions, and the input parameters. For problems of the complexity addressed in repository modelling, existing databases are often not

---

[108] The Royal Society, 1994, *op. cit.*, pp. 69-70.

adequate. These can be improved by measurement where possible, extrapolation from existing data, or from assessing the judgements of experts in eliciting likely parameters. Each approach has its advantages and disadvantages, and all three have been used.[109]

- In systems whose characteristics may evolve over time, a specification of these possible changes, together with an estimate of probability.
- In systems - such as here - where there are inherent uncertainties, some way of dealing with these. One approach is to run a model simulation under different conditions (e.g. different assumptions of the changes that may occur over time; sampling starting parameters from a distribution of those thought to be reasonable) and produce a large number of possible outcomes, possibly with probabilities attached. A partially useful analogy would be the running of weather forecasting programs many times with different reasonable inputs to see how the resulting forecast may vary depending on the initial assumptions. This can help to set bounds on likely future behaviour within the validity of the model.

A further important aspect of modelling is validation: it needs to be run in well-understood systems to demonstrate that it produces known results. Validation can be against, for example, the results of laboratory experiments, field tests and observations. In problems involving very long times, natural analogues may have a particularly useful role to play in model validation.

A related approach is based on "*scenarios*", which are defined as broad-brush descriptions of the initial state of a system and its future evolution.[110] For example, in considering ground water transport, we might consider a climatic transition through a colder one to glaciation, the resulting effects of which would be likely to include changes in rainfall and sea levels which would influence groundwater flow. The formation and melting of an ice sheet over a repository could also affect groundwater flow, and also possibly trigger earthquakes which could open up cracks, or alter the stress state within the rock, with further consequences for groundwater flow.

For each scenario, more detailed descriptions are generated that are sufficiently specific for post-closure performance assessment calculations. In principle, enough scenarios would be constructed to cover all possibilities of the site's evolution, assigning a probability to each. In practice, however, constructing a full set of self-consistent scenarios is not straightforward and, to date, it would appear that work has been restricted to a small number of restricted scenarios to explore limited aspects of a site's possible evolution.[111]

There is disagreement as to which approach is preferable. One might argue mathematically-based modelling to be potentially more rigorous, less uncertain, and able to provide a framework for reasoned discussion between practitioners. However, effective modelling also requires the specification of a full set of self-consistent possible future changes to the system being modelled.

---

[109] The Royal Society, 1994, *op. cit.*, chapter 5.

[110] OECD Nuclear Energy Agency, 1992, *Systematic approaches to scenario development.*

[111] See, for example, UK Nirex Ltd, 1993, *op. cit.*

I now consider a few examples relating to modelling some of the processes involved, in an attempt to assess both the present state of repository modelling and its potential for future development. It is useful to start this from the near-field of waste container, surroundings, and the immediate environment of the repository.

3.2 The near-field

As the repository itself is the part of the disposal system over which we have the most control, we might expect this to be the best-understood part of the overall near-field/geosphere/biosphere system. There are however a number of complexities, particularly relating to the chemistry, as exampled below.

The engineering of possible repositories varies, but there are perhaps common aspects on which a brief discussion may be based. *First,* the immediate containment of the waste in drums or boxes. These may be either an integral part of the system designed to retain the contents against corrosion for a significant time (e.g. Swiss proposals include 25 cm thick stainless steel overpacks to give long-lived containment of vitrified waste[112]), and other, more corrosion-resistant materials have been considered, e.g. titanium, copper, and a copper-steel composite.[113] Alternatively as in the UK situation, the drums may be assumed to corrode relatively rapidly, and therefore not be considered part of the containment. *Secondly*, the containers may be surrounded by an alkaline grout to keep the pH high to reduce the solubility of most (though not all) of the radionuclides, and/or a material such as bentonite clay[114], with the intention of (a) suppressing groundwater flow through the wastes, and (b) retarding radionuclide movement when eventually they are released to groundwater. There is some disagreement as to the advisability of a bentonite layer, there being arguments that it could lengthen the time to establish the high pH regime desired to limit solubility, could degrade in contact with cementitious backfill, and could prevent gas escape.[115] *Thirdly*, assumptions made about the post-closure integrity of the immediately surrounding repository cavity will have a bearing on the eventual uptake by the geosphere of material from the store.

We need to both (a) understand, and (b) be able to model effectively, the post-closure evolution of this effectively two-barrier system: the physical barrier of the containers which initially both retains the waste and prevents groundwater entry, and the chemical barrier of the surrounding backfill. Considering the second barrier only, we can identify some fairly complicated chemistry that needs to be well understood. For example:

---

[112] The Royal Society, 1994, *op. cit.*, p.45.

[113] UK Parliamentary Office of Science and Technology, 1997, *op. cit.*, p. 55, and The Royal Society, 1994, *op. cit.*, p. 45 and p. 49.

[114] Swedish Nuclear Fuel and Waste Management Co. (SKB), *Final disposal of spent nuclear fuel, the importance of the bedrock for safety*, 1992, Technical Report 92-20, Stockholm.

[115] UK Nirex, 1994, quoted in The Royal Society, 1994, *op. cit.*, p. 118.

- the designed high pH regime is expected to reduce the solubility of most of the radionuclides involved. However, the chemistry of different radionuclides can be very different, and so the effect on solubility will be nuclide specific. The long-lived ^{36}Cl and ^{129}I are particular high solubility cases in point. More generally, to be able to model the effect of the high pH regime on solubility, a good database is needed. Though a significant amount of data exists, and international collaborations have enhanced it[116], the database may still be inadequate.
- Once groundwater enters the repository, its reaction with any cementitious backfill is likely to lead to changes in buffering activity and capacity. Although the buffering capacity is expected to be high, the time over which it will remain effective will depend on the throughput of ground water to the geosphere. A possible problem is that cement hydrates are metastable with respect to silicate minerals, and if these latter are formed, they would form less effective buffers, although preliminary experiments suggest buffering capacity may not be lost.[117]
- The barrier itself may be modified by incoming groundwater. For example, (a) $CaCO_3$ precipitation may reduce permeability and (b) reaction with CO_2 derived from organic degradation may reduce buffering capacity.
- The primary containment will begin to corrode once saturated with groundwater. Corrosion is a complex process that will be influenced by the oxygen content of the surrounding water and the presence of ions such as Cl^-, the influence of which has been studied.[118] Anaerobic corrosion of steel will produce solid products and hydrogen gas which, together with methane and CO_2 from organic matter degradation, could affect groundwater flow and be a potential carrier gas for some radionuclides. There is also evidence that polar organic molecules resulting from both inorganic and microbial degradation may enhance radionuclide solubility and hence enhance mobility. Experiments on this kind of problem are not easy, though work has been done.[119]
- In addition to solubility, release of radionuclides into groundwater is influenced by their possible sorption onto the backfill. Although a sound thermodynamic framework exists, the necessary thermodynamic data are not all available. Furthermore, there are problems of kinetics which are difficult and imperfectly understood. Sorption of radionuclides is also important in considering radionuclide transfer across the interface with the geosphere. This problem raises further issues about which there is considerable uncertainty: for example, there is a significant lack of understanding of the mechanism of the sorption for many nuclides, and the complexity of the system implies it may not be possible to use sorption parameters in any predictive sense.[120]

---

[116] e.g. the so-called CHEMVAL project sponsored through the Commission of the European Community (CEC).

[117] A. Atkinson, S.J. Williams, and S.J. Wisbey, 1993, NSARP reference document, *The near-field*, Nirex safety studies report, NSS/G117.

[118] C.C. Naish *et al.*, 1993, *The anaerobic corrosion of carbon steel in concrete*, Nirex safety studies report, NSS/R273.

[119] C.C. Naish *et al.*, 1993, *op. cit.* P.J. Agg *et al.*, 1993, NSARP reference document, *Gas generation and migration*, Nirex safety studies report, NSS/G120.

[120] The Royal Society, 1994, *op. cit.*, p.124.

- A major problem in predicting radionuclide transport is the very long time scale involved. Laboratory experiments can only be undertaken over short times. Performing experiments at higher temperatures to simulate longer times is problematical in systems where the phases obtained at these higher temperatures are not those that would exist under repository conditions. Here the study of natural analogues may be particularly helpful.

The above points illustrate the complexity of the problem within the near-field zone. The models we use to examine near-field behaviour must be able to handle this complexity. Moreover, there are deficiencies in our understanding of some of these processes (aqueous solution chemistry even of familiar ions and simple molecules is far from being understood, while the influence of ions on solubility of other molecules is even less well understood).[121] A great deal of work has been done to both (a) improve our understanding of the processes involved, and (b) fill in the gaps in the databases required for reliable modelling. However, work still seems to be needed in many areas, two examples being the organic matter enhancement of solubility of certain radionuclides, and the evolution and migration of gas.[122] There appears not yet to be an adequate way of combined modelling of water flow, the generation and movement of gas, and heat flow that is needed to understand the effects of gas generation. Overall, although the near-field environment is that over which we have the most control, the complexity of the many processes likely to be involved implies that further significant development is needed before we can be sufficiently confident of modelling this environment reliably.

3.3 The geosphere

A - if not the - main question here relates to the flow of groundwater which will control the rate of radionuclide transport. Liquid flow modelling is relatively well developed in examining flow through porous media and through fractures systems. It is, however, computationally demanding, even more so when, as in the case of modelling likely radionuclide migration over very long time scales where significant uncertainties are involved, an ensemble of calculations need to be performed in order to explore the range of possibilities that need to be considered. Thus there have been attempts to use simplified programs, for example by applying continuous media models of what are in reality non-continuous ones[123], and even to work in a much simpler two-dimensional framework rather than make full three dimensional calculations.[124]

With continuing increases in computational power, there have been major improvements in modelling systems in general. With expected further improvements,

---

[121] See for example F. Hofmeister, 1988, *Arch. Exp. Pathol. Pharmakol.*, vol. 24, p.247; A. Ben-Naim, 1980, *Hydrophobic Interactions*, Plenum, New York; W. Blokzijl and J.B.F.N. Engberts, 1993, *Angew. Chem.*, vol. 32, p.1545; J.L. Finney and A.K. Soper, 1993, *Chem. Soc. Revs*, vol. 23, p.1.

[122] The Royal Society, 1994, *op. cit.*, pp. 124-5.

[123] *Ibid.*, p. 132.

[124] See for example *ibid.*, p. 102.

many of the computational restrictions that have driven towards the use of simplified models are being lessened. It seems reasonable to expect that available computational resources will enable us to deal adequately with the technical aspects of the groundwater flow modelling required, thus countering some of the serious criticisms that have been made of some simplified approaches.[125]

A further problem with effectively modelling radionuclide transport through the geosphere concerns the validity and reliability of the input data. To characterise the geosphere we need a good knowledge of the structure of the fractured system through which the groundwater will flow. The nature of these fractures with respect to the flow properties through them (such as their extent, spacing, orientation and conductivity) needs to be specified. A certain amount of data is obtained through borehole cores. These data are inherently limited: the lateral extent of the observed fractures cannot be determined from core observations alone. Furthermore, we cannot continue to drill further boreholes to improve the database, as this would eventually begin to have a major effect on the flow properties of the system. Although supplementary data to assist understanding the extent and connectivity of a fracture system can be obtained by indirect means (e.g. local well flows, pressure tests), these methods also have significant limitations. Use can be made of established statistical techniques to obtain needed parameters by extrapolation of field data.[126]

Provided the fracture system can be adequately specified, effective modelling of groundwater flow under present conditions is likely to be possible. We need however to consider also how *changes* to the geosphere that may occur in the next million or so years could affect the behaviour of the model. Earthquakes do occur away from plate boundaries (possibly associated with icecap unloading), and the effects of these on groundwater flow are thought to be subtle and difficult to predict.[127] There is active research in this area, and apparently no consensus as yet, though it seems to be accepted that what was originally an impermeable seal may become a flow channel after a disturbance.[128] Entry into a glacial period - highly likely in the UK in the next million years if we take notice of the historical record - will lead to, among other possible consequences, a reduction in sea level, and an increase in lithosphere stress through ice loading. In addition to possibly changing the patterns and rates of groundwater recharge and discharge, these changed stress conditions have the potential to change groundwater flow. The possible extent of such changes appears not to have been extensively discussed. The effect will depend upon the interaction of the perturbation with the existing stress field, an interaction which is imperfectly known and actively debated.[129] There also remains the possibility of other long-time perturbations that - unlike earthquakes and glaciation - are not present in the historical record. We cannot rule out the possible influence of man-made perturbations, for example the loss of the Greenland ice sheet.

---

[125] See for example *ibid.*

[126] See for example *ibid.*

[127] *Ibid.*, p. 155.

[128] *Ibid.*, p. 156.

[129] *Ibid.*, p. 157.

To include the effects of future possible events on modelling changes in groundwater flows will require calculations using a large number of input parameters. This becomes increasingly difficult as the time period to be covered increases. Validation of the modelling may also not be possible.[130]

Further points that relate to effective modelling of radionuclide transport in the geosphere include the following.

- As heat flow, hydrological transport, and mechanical stress can interact with one another, a more complete approach to modelling would couple all these effects.[131] These models are even more complex, and it seems more development will be needed before they can be effectively used in calculations on candidate repositories. As with the simpler modelling approach discussed previously, problems here also include obtaining (a) an adequate geological description and (b) the input data required.
- Migration of gases, either dissolved in groundwater or as a separate phase, has significant implications for groundwater movement. The largest - but by no means the only significant - uncertainty here relates to the modelling of two-phase (gas/water) flow through a fracture system. Major developments in understanding are needed here.[132]
- Radionuclide retardation by sorption on solid surfaces was mentioned in the discussion of the near-field system. Related effects are also expected to play a major role in controlling the migration of radionuclides through the geosphere. Prediction of the effects of sorption is far from straightforward, particularly in fractured rock. Gathering the necessary data is also not straightforward - the lowest diffusion coefficients are the most difficult to measure. Also to be considered are competition from organic complexes, the effects on radionuclide transport of association with colloids, and the effects of the chemically-disturbed zone around the near field which may act to modify sorption.

In looking at the geosphere, we again meet problems of considerable complexity. Although modelling techniques are likely to develop even further with enhanced computational power, basic problems remain in areas such as characterising the rock fracture system adequately for such modelling to give reliable results, dealing with the perturbing effects of gas on radionuclide migration, and understanding the extent to which sorption can retard radionuclide migration. When we look forward in time, adjusting our models to take account of possible changes that are likely to affect significantly the groundwater flow is even more problematical. There seem to be topics in the geosphere system for which present knowledge is not sufficient to fully resolve the relevant problems.[133]

---

[130] N. Oreskes *et al.*, 1994, "Verification, validation, and confirmation of numerical models in the earth sciences", *Science*, Vol. 263, p. 641.

[131] L. Jing *et al.*, 1993, *DECOVALEX - mathematical models for coupled T-H-M processes for nuclear waste repositories*, Report of phase 1, SKI technical report 93:31, Stockholm.

[132] The Royal Society, 1994, *op. cit.*, p. 144.

[133] *Ibid.*, p. 148.

3.4 The biosphere

Biosphere modelling has a relatively successful history. Its validation should also be more straightforward than geosphere modelling. Thus, we might expect relevant processes such as the distribution of radionuclides in the surface environment and their concentration in organisms to be amenable to reliable modelling. Similarly, adequate dose calculations for humans and other organisms might be expected to be reasonably achievable.

Compartment models divide the biosphere into local, regional, and global compartments, with each compartment being defined by the masses of solid material (e.g. soil and sediment) and water within that volume. Transfer rates between compartments are determined, based on the physical processes involved in such transfers. Assumptions are made that radionuclide activity entering a compartment is instantaneously distributed throughout its volume. The mathematics of these models is straightforward.

In adapting these models to post-closure performance assessments of repositories, there may be issues which need some attention. For example, the interface between the geosphere and biosphere compartments seems not to have been addressed, the assumption generally being that the output from the geosphere transport model is input directly into the compartment thought most appropriate.[134] There is also no consensus as to whether single values of relevant parameters derived from models (e.g. dose per unit intake in assessing risks to individuals) should be used, rather than making an "ensemble" of calculations with sample values taken from a probability distribution of parameters.[135] This approach has been justified by making the argument that these values are considerably more certain than are many parameters in near-field and geosphere models.[136]

A further problem - similar to the treatment of the geosphere - is the inclusion of changing biosphere conditions in these calculations. Climatic change over the time period we have to consider is a realistic possibility, and one which would have a major effect on the way the biosphere is modelled.[137]

Overall, biosphere modelling seems to be relatively well advanced, compared to that of the near-field and geosphere. Uncertainties in treating the present day system seem to be smaller than in these other two zones. Nevertheless, increased attention needs to be given to handling the influence of possible climatic and other perturbations that may affect the biosphere over periods of up to about 10^7 years through the development of dynamic biosphere models. Attention to how best to deal with the interface between the biosphere and geosphere would result in improved modelling of the overall problem. A consensus also needs to be reached on the use of single parameter values in aspects of the

---

[134] *Ibid.*

[135] *Ibid.*, p. 152.

[136] *Ibid.*, pp. 149-50.

[137] C.M. Goodess, J.P. Palutikov and T.D. Davies, 1992, *The nature and causes of climate change: assessing the long-term future*, Belhaven Press, London.

modelling as against the more expensive, but inherently more valid, sampling of parameter values from probability distributions.

3.5 Summary: the state of repository modelling

Modelling the dispersion of deposited radionuclides over a timescale up to 10^7 years is not straightforward. Increasing computer power means that more sophisticated models are likely to be useable, though there may be intrinsic problems in model validation. Many of the processes involved are complex, involving many interacting components. Our understanding of many of these remains imperfect. Furthermore, our ability to characterise sufficiently well some of the systems concerned - e.g. the flow characteristics of the geosphere surrounding a repository - is limited. The extensive databases we need as input parameters to modelling calculations are large, and significant gaps remain which need to be plugged reliably. Our ability to handle the effects of possible future events remains problematical. In the face of these difficulties, much impressive work has and is taking place to address these problems. However, on the evidence presently available, we do not yet seem to have consensus solutions to the major outstanding problems.

As a note of caution, however, we should comment that certain environments raise fewer difficulties. For example, a dry storage repository (e.g. in salt formations) - provided we can be adequately sure there will be e.g. no future flooding, or brine inclusions[138] - does not have the groundwater flow problem around which many of the difficulties revolve. However, for nuclear power to be a continuing power production method world wide, if ILW and HLW disposal cannot be avoided (e.g. by using other fuel cycles) the repository capacity that will be needed is likely to require construction in groundwater situations.

4. Utilisation in Power Reactors

Plutonium can be mixed with uranium oxide to produce a mixed $(Pu,U)O_2$ MOX fuel that can be utilised in some existing reactors. Technically, it would be possible to use this method to give a net burnup of plutonium, while extracting some of the energy that would be lost in the deep disposal route.[139] New light water reactors designed for MOX fuel could be built to give more efficient burnups.

The MOX option deals only with the Pu component of the waste stream. However, the Canadian Deuterium Uranium reactor (CANDU) has a higher rate of Pu consumption, and overall consumption of recovered plutonium, uranium, and the minor actinides is possible.

Other possibilities of burning plutonium include (a) fast neutron reactors which could be used in a mode that almost completely burns up the plutonium, and could also

---

[138] The Royal Society, 1994, *op. cit.*, chapter 5.

[139] The Royal Society, 1998, *op. cit.*

minimise the production of minor actinides; and (b) the helium-cooled high temperature gas-cooled reactor, which can eliminate 90 % of the initial plutonium charge.[140] Though the prospects for both these systems in the West are presently not good, technically they are options that merit consideration.

Apart from the CANDU reactor, the above options in the waste context relate largely to plutonium, rather than the full spectrum of HLW and long-lived ILW. However, in the context of an expanded nuclear power programme, these options have the advantage of reducing the waste problem by producing less hazardous waste in the first instance. There would seem to be significant potential for reducing the waste problem through alternative fuels and reactor designs. In addition, the Pu production - and the consequent other waste components - could be reduced by using a matrix other than uranium. Those that have been tested include thoria, silicon carbide, and spinel. Though extensive work, including tests of fuel performance and safety analyses, would be needed, the potential of these options to reduce the severity of the waste problem warrants further consideration.

5. Transmutation and Sub-Critical Reactor Systems

There are influential views that neither partial conversion of plutonium, nor disposal of this and other long-lived active waste, should be regarded as an optimal long-term solution to either the waste problem or a future expanded nuclear power programme. Ideally, a long-term strategy should aim to close the fuel cycle, even if the problems of long-term disposal are ultimately solved convincingly. An option that could possibly begin to do this, as well as possibly deal with the most difficult parts of the present waste legacy, is accelerator-driven subcritical reactors.

Work on such systems, which are driven by neutrons produced by high energy protons hitting an appropriate target, has been proceeding for over two decades, particularly in Japan and the US.[141] It received a fillip from work on high-power accelerators as part of the Star Wars programme, and much progress has been made subsequent to the cancellation of that programme. Over recent years, the studies of ATW (accelerator transmutation of waste) appear to have developed sufficiently that they warrant consideration as options for both (a) disposal of the waste that will be built up by the present nuclear power programme, and (b) the basis of a future nuclear power programme that avoids some of the difficulties associated with thermal reactors. The latter aspect will be dealt with in Chapter 16, by van der Zwaan. I consider briefly here the potential for waste disposal through transmutation.

The ATW system consists of a high power proton accelerator (40 MW in the current US study[142]) driving one or more subcritical burners. Each of these burners consists of a spallation target to produce the neutrons to be used in the burnup processes. The current US study has chosen lead-bismuth eutectic (LBE), which can act as both

---

[140] *Ibid.*

[141] See, for example, C.D. Bowman *et al.*, 1992, *Nuclear Instr. and Methods*, vol. A230, p.336.

[142] F. Venneri *et al.*, 1998, 6th International Conference on Nuclear Engineering, ASME.

spallation target and coolant for a number of technical reasons.[143] There is extensive Russian experience with LBE, which the US work builds upon.[144]

The system is claimed to be able to effectively destroy all transuranics including plutonium, together with the longer-lived fission products, in particular Tc and I which are particularly troublesome with respect to disposal. Burnups of over 99.9 % are claimed for all these elements.[145] The system is argued to be ideally suited to the incineration of material that is (a) not well characterised, (b) burns poorly or not at all in reactors, and (c) has potentially unstable and hazardous reactivity responses (e.g. Np, Am, Cm).

What is not passed through the ATW burners is mainly short-lived fission products such as ^{90}Sr and ^{137}Cs, for which disposal is still required. However, as most of the radionuclides that the system cannot destroy efficiently are "short-lived", most radioactivity will decay over a period of a few hundred, rather than the 10^7 years required otherwise (activity for a 30 year half-life is down to 0.1 % original activity after 300 years). The technical problems of storage - in addition to the volume of waste to be dealt with - are thus very considerably reduced.

Pyrotechnical processes would be used to separate off the transuranics and long-lived fission products from the shorter lifetime material that is destined for storage. These separation processes are argued to be less expensive than traditional aqueous chemistry techniques, and to produce less consequential processing waste.[146]

Subcritical reactors such as these could be considered either (a) to destroy the inventory of long-lived isotopes produced by the existing nuclear power programme, or (b) as an option for a future nuclear power programme that would have a much reduced long-term waste problem. In the case of (a), the operation of a number of burners would be phased in to deal with the accumulated material. As each is decommissioned, the resulting contaminated material can then be burned up in remaining systems until only one remains. This would then leave a relatively small amount of material for final long-term disposal. Option (b) is outside the terms of reference of this chapter, except to note the implications of a less-problematical waste stream.

There are perhaps still some unanswered questions concerning the viability of the ATW option, and work still needs to be done on it. However, this is also the case for the long-term disposal option considered earlier: one considered view is that the scientific programmes required can be expected to take decades to complete.[147] It might be commented that the science of the ATW route is likely to be inherently less complex (it relates to 'hard' physics that can be tackled with less controversy) and involves fewer potential uncertainties than does the disposal route. The fact that it is a potential power source also adds to its attraction. Although some earlier statements on the ATW route

[143] F. Venneri *et al.*, 1998, *Accelerator-driven transmutation of waste (ATW)*, Technical Review at MIT, Los Alamos National Laboratory report LA-UR-98-608.

[144] Communications and contract reports from the Inst. of Physics and Power Engineering (Obninsk) and EDO-Gidopress (Podolsk), Russia. Quoted in F. Venneri *et al.*, 1998, *ATW, op. cit.*

[145] F. Venneri *et al.*, 1998, *op. cit.*

[146] M.A. Williamson, 1997, "Chemistry technology base and fuel cycle of the Los Alamos accelerator-driven transmutation system", *Procs. of Global '97*, p. 263.

[147] The Royal Society, 1998, *Submission to the House of Lords Select Committee on Science and Technology inquiry into The Management of Nuclear Waste*, p.1.

have been negative[148], these do not take account of the more recent work. There seems to be a strong case to look at this option very seriously.

[148] See, for example, UK Radioactive Waste Management Committee, 1995, RWMAC Forward Look, "Comparisons of accelerator-based with reactor-based nuclear waste transmutation schemes", *Progressions in Nuclear Energy*, July 7, 1994.

Chapter 10 Spent Fuel Management

by
Pierre Goldschmidt*

Introduction

How best to manage spent nuclear fuel has been one of the most controversial issues world-wide for more than twenty years. The issue raises essentially three basic questions:

1. Should spent nuclear fuel (which will contain some 95 % uranium and 1 % plutonium) be considered as a waste, and, after being stored during a few decades, be encapsulated and permanently disposed of in an appropriate geologic underground formation? This is the so-called "once-through" fuel cycle option (see the previous Chapter, by Finney).

2. Should spent nuclear fuel be considered as containing valuable materials for electricity production that should be extracted through reprocessing and recycled in fast breeder reactors (FBRs) or possibly in light water reactors (LWRs)? This is the "reprocessing/recycling" option.

3. Should each country operating nuclear power plants be responsible for the final disposal of its encapsulated spent nuclear fuel or nuclear wastes in an appropriate geologic underground formation within its own borders, or is it preferable to manage nuclear wastes on an international or regional basis?

The answers to the first two questions depend first of all on the perspective one has about the future development of nuclear energy to produce electricity in the long-term. Clearly, if one believes that the recourse to nuclear power plants for electricity production will inevitably be phased out, say by the middle of next century, then reprocessing spent fuel to recover its remaining energy content is obviously not a necessity. If, on the other hand, one believes that nuclear power plants must and will constitute a significant portion (for instance 30 percent) of the long-term electricity production mix, then it is most probable that fast breeder reactors producing more plutonium fuel[149] than they consume will be needed and this, in turn, will require reprocessing spent fuel to extract its plutonium content.

* Pierre Goldschmidt is General Manager of SYNATOM, responsible for the nuclear fuel cycle management of Belgium's nuclear power plants. The views expressed in this chapter are his own.

[149] Or U-233 fuel, in a thorium fuel cycle.

The other major question that divides experts on whether or not to reprocess spent fuel is their different judgement on which solution is preferable from a non-proliferation point of view. We will, in the following paragraphs, address these basic aspects of the issue.

1. The Energy Framework and Breeders

In a 1993 paper on "Global Energy and Electricity Futures", Dr. Chauncey Starr, President Emeritus of the Electric Power Research Institute, stated: "By the middle of the next century, global energy demand driven by population and economic growth, will be in the range of 2-4 times the present level, depending on the effectiveness of energy efficiency and conservation globally. Even with maximum realistic conservation, the electricity component will be more than 4 times present usage. A massive expansion of non-fossil sources would be needed to slow the future annual increase in carbon dioxide to the atmosphere. We cannot accept and should not wish the developing nations to forego the benefits of abundant energy that the industrialised world has enjoyed for many decades."[150] For a description of future global energy demands and the need for the expansion of the use of non-fossil energy resources (see Chapter 2, by Fetter).

In a report published in 1995, an American Nuclear Society (ANS) "Special Panel on Protection and Management of Plutonium" strongly supports the development and use of renewable energy sources, which represent varying ways of utilising solar energy, but emphasises that technological and economic factors, as well as solar energy's limited availability in many regions, will keep its contribution limited. The ANS Panel, "therefore, shares the view of energy experts that nuclear energy will play an increasing role in meeting world energy demand in the next century. This does not mean that other energy sources should be overlooked."[151]

"Today's commercial power reactors derive no more than 1 % of the energy value of the uranium required for their operation. There is little, if any, dispute, even on the part of strong opponents of plutonium recycling, that, if nuclear power is to make a major, long term contribution to meeting world energy demand, this level of resource utilisation must be increased dramatically through the use of breeders or other advanced reactors. The issue is when this is likely to be required.

It is not surprising that given its size and diversity, our Panel held a range of views on this question. Indeed, the uncertainties inherent in energy forecasts make it impossible to predict this date with any assurance. What we do know with certainty is that the time scales for development are very long, and the costs of losing continuity are very high. Accordingly, and this is the Panel's central finding in respect to the future need for plutonium, we were unanimous in the view that it is unnecessary to be certain of precisely when the breeder is needed to conclude that it would be imprudent in the extreme not to continue breeder development, and to do so at something close to the

---

[150] American Nuclear Society (ANS), Special Panel Report, *Protection and Management of Plutonium*, August 1995, Executive Summary, p. 11.

[151] ANS Special Panel Report, *ibid.*, p. 36.

current level. It is, of course, inherent in this view that there is a reasonable possibility that the breeder will be needed to meet energy demand in an environmentally acceptable way by sometime around the middle of the next century, since it would be difficult to rationalise continued development if it could be determined with certainty that the time horizon is considerably more distant than this.

Some may argue that the discovery of large additional uranium resources is sufficiently certain that we know now that the breeder will not be needed until well into the latter part of the next century. They may be right; they may also be wrong.

In short, our Panel concluded that the risks to our collective ability to meet world energy demand, that could arise from an interruption of breeder development, far outweigh any risks of proceeding with a constrained and responsible development effort conducted under the effective controls of the nonproliferation regime. The Panel took particular note of the fact that future energy growth will take place primarily in the developing world, and we cautioned against policies that would have the effect of depriving these countries of the benefits of abundant energy that the industrialised world has enjoyed for so long."[152]

Therefore, until plutonium is needed as breeder fuel, spent fuel should either be stored in a retrievable manner or reprocessed. However, when reprocessing takes place, plutonium accumulation should be avoided by prompt recycling in existing commercial reactors.

2. The Non-proliferation Issue

2.1 A US perspective

The policy of the United States with regard to the reprocessing of spent nuclear fuel has varied widely, from the initial optimism which existed during the first years of the nuclear era to the strong opposition of the Carter administration in 1977, and to the present acquiescence in the recourse to reprocessing in those industrialised countries which already have very advanced nuclear programs. President Clinton has made clear his antipathy to reprocessing in the United States while at the same time professing no desire to interfere with reprocessing in Western Europe and Japan: "The United States does not encourage the civil use of plutonium and, accordingly, does not itself engage in plutonium reprocessing for either nuclear power or nuclear explosive purposes. The United States, however, will maintain its existing commitments regarding the use of plutonium in civil nuclear programs in Western Europe and Japan."[153]

The US opposition to reprocessing and recycling is essentially based on the fear that, if it took place worldwide, it would increase significantly the risk of proliferation. Already in his Statement on Nuclear Policy on October 28, 1976, President Ford stated:

---

[152] "The American Nuclear Society Special Panel on Protection and Management of Plutonium: A Review", M.B. Kratzer, August 1995.

[153] Nuclear Proliferation Factbook, Committee Print, 103d Congress, 2d Session, S. PRT 103-111, p. 195.

"I have concluded that the reprocessing and recycling of plutonium should not proceed unless there is sound reason to conclude that the world community can effectively overcome the associated risks of proliferation. I believe that avoidance of proliferation must take precedence over economic interests."

Similar fears have been expressed over the last twenty years by a number of experts and organisations, usually those opposing the use of nuclear energy. In November 1997, the "Citizens' Nuclear Information Center" published a report[154] which stresses, *inter alia*, that in the case of the reprocessing/recycling fuel cycle, present safeguards are not an appropriate guarantee. They have calculated that, in a large reprocessing plant (800 ton/y), the minimum amount of diverted plutonium which could theoretically be detected with a probability of 95 % is about 220 kg/y, enough to produce 6 to 10 nuclear explosive devices. They also point to the fact that in 1994 it was found that about 70 kg of plutonium were held-up (stuck to the surface) in the remote-handling equipment at the Tokai Mixed Oxide fuel (MOX) fabrication plant in Japan. The same experts also believe that existing physical protection concepts, which for obvious security reasons are not in the public domain, are defeatable, and that plutonium and MOX transports are particularly vulnerable to theft or diversion by highly trained terrorist organisations.

However, over the last few years, an increasing number of members of the US Congress and the scientific community have been taking an opposite view. In 1996, Gregg Renkes, Staff Director of the US Senate Committee on Energy and Natural Resources, stated that:

"US non-proliferation policy is too focused on the risks associated with reprocessing and not concerned enough with the real proliferation risk of growing inventories of plutonium worldwide. US nuclear policy has been blind to the basic fact that you don't eliminate plutonium by burying spent fuel underground, the current sole objective of our nuclear waste policy. Only burning plutonium in reactors can eliminate it. We have failed to account for the long-term proliferation risks associated with the massive build up of spent fuel in storage facilities from which plutonium could be retrieved.

Further, US non-proliferation policy is not having an impact on nuclear programs in other nations. Only the US and a few other countries have, as a matter of policy, rejected plutonium use. The rest of the world will not turn away from plutonium as an energy source. Reprocessing is an international fact; the US policy has simply not worked. What is worse is that reduced involvement in the technology reduces the impact the US can have on international control regimes and non-proliferation technology development. (...) "[155]

"Despite current US non-proliferation policy, the successes experienced by other countries who employ closed fuel cycle technologies and the failure of our own waste

---

[154] "Comprehensive Social Impact Assessment of MOX use in LWRs", IMA Project, Citizens' Nuclear Information Center, November 1997.

[155] US High-level Waste Management Policy and the Reprocessing Option, Gregg D. Renkes, ANS/ENS International Meeting, Washington, 10-14 November 1996, p. 4.

disposal program are driving a reexamination of the utilisation of reprocessing technologies in the US. There are no US laws that ban reprocessing of US spent fuel."[156]

"The back-end of our fuel cycle is broken. We have traded reprocessing, fast reactors, and the development of rational and reasonable disposal techniques for a failing permanent disposal program."[157]

For Nobel Laureate Glenn Seaborg, "Proliferation to date has not resulted from the diversion of material from peaceful uses, and I believe it is clear that, while we must remain alert to this possibility, the fuel cycle is not the principal or most likely threat of proliferation at present or in the future. The main threat has been and will continue to be the development by countries of small, dedicated facilities to produce either plutonium or highly enriched uranium deliberately for weapons purposes. This route to proliferation will always be with us, even if nuclear power is completely phased out.

We cannot rely on the difficulty of reprocessing as an effective barrier to proliferation any more than we can rely on the difficulty of fabricating a nuclear weapon by those who come into possession of enough fissionable material.

Burying spent fuel is not a solution to the proliferation risks of plutonium. It simply transfers the risks to future generations. We must find ways of making it clear to policy-makers and opinion-leaders alike that those who advocate the disposal of spent fuel, no matter how well-intentioned, do not necessarily occupy the "high ground" in the non-proliferation debate."[158]

Similar views were expressed by the members of the ANS Special Panel. The Panel also concluded that the number of fuel cycle facilities handling separated plutonium should be limited and that this will certainly call for a regional approach.

"Consideration of the retrievability of spent fuel from repositories leads to a conclusion which is explicitly stated in the report and is reflected as well in the Sub-Panel Report on nuclear fuel cycles. This is, that from the standpoint of non-proliferation considerations alone, fuel cycles in which plutonium accumulation is limited through recycle under effective safeguards and controls can in the long run be a better non-proliferation policy than fuel cycles that avoid reprocessing and recycling but result in the indefinite accumulation of plutonium in spent fuel. If we look at the global or even a national nuclear situation at a given point in time, plutonium present in separated form will, of course, pose greater proliferation risks than spent fuel, but if we look at the situation into the future, a world, or even a particular country, in which spent fuel has accumulated for, say, 50 years will obviously pose greater proliferation risks than one in which spent fuel accumulation has been avoided or limited by recycling."[159]

---

[156] *Ibid.*, p. 3.
[157] *Ibid.*, p. 1.
[158] Nuclear Recycling, February 1996, N°2.
[159] ANS, Special Panel Report, *op. cit.*, Kratzer, *op. cit.*

2.2 A European perspective

2.2.1 Historical background

In Europe, the attitude towards spent fuel reprocessing has followed a quite different trend. Within a year after President Carter's decision to ban reprocessing, and the failure of the International Nuclear Fuel Cycle Evaluation he had initiated to demonstrate that the once-through cycle was superior to the reprocessing/recycling fuel cycle, most of the large European and Japanese nuclear utilities operating light water reactors (except Spain) signed long-term reprocessing contracts on a cost plus fee basis with COGEMA (*Compagnie Générale des Matières Nucléaires*) in France and BNFL (British Nuclear Fuel Ltd.) in the United Kingdom.

As recalled in Chapter 13 (§ 4.1), reprocessing was at that time the only spent fuel management option authorised under the German Atomic Law. In Belgium, after a thorough analysis of the different options by an independent "Blue Ribbon Panel", reprocessing became in 1976 the reference solution, a policy adopted by the Government in 1979 and confirmed by the Parliament in 1982 and 1983. The first departure from the reprocessing policy in Europe came from Austria, after the decision in 1978 not to start-up the Tullnerfeld reactor, and from Sweden, in the wake of its 1980 referendum in favour of phasing out the operation of all nuclear power plants by 2010. Thereafter, between 1985 and 1990, Swedish utilities were able to sell most of their reprocessing commitments to German and Japanese utilities.

It is, however, worth mentioning that reprocessing of 140 tons of Swedish spent fuel from Oskarshamn-1 and -2, shipped to Sellafield (UK) from 1969 to 1982, has started in July 1998. It is expected to yield 136 tons of uranium and 900 kilograms of plutonium. Oskarshamn will seek to use this plutonium as MOX-fuel at the plant.

A new blow to the reprocessing industry was the termination in 1989 of all construction work on the German Wackersdorf reprocessing plant due to public opposition and ever more complex licensing procedures. Thereafter, the German Atomic Law was modified in 1994 to allow the once-through fuel cycle as an additional valid spent fuel management method.

In the meantime, in Belgium, the reprocessing policy had been questioned by a Parliamentary Commission for the first time in 1990. At the end of 1993, after a one year long open debate in the Belgian Parliament, the Government decided that the reprocessing/recycling option and the once-through fuel cycle option should, from then on, be considered, studied and developed on an equal footing.

In France, until very recently, the only spent fuel management policy was to assume that all spent fuels would be reprocessed. In 1998, two Members of the French Parliament submitted a "Report on the back-end of the nuclear fuel cycle" to the "*Office Parlementaire d'Evaluation des Choix Scientifiques et Technologiques*". This report concludes that, for economic reasons, about one third of the total spent fuel produced in the country should not be reprocessed in the foreseeable future and that the back-end of the French fuel cycle should now be on a dual track.[160]

In the Netherlands, however, reprocessing remains the only spent fuel management method in force. In 1978, Dutch utilities signed reprocessing contracts with COGEMA, for the Borssele nuclear power plant, and with BNFL, for the Dodewaard plant. In 1993, a new contract was signed with COGEMA to cover all the reprocessing needs of Borssele until the end of its life. As recently as June 1997, the Dutch Minister of Economic Affairs, G.J. Wijers, wrote in a letter to the Chairman of the Lower Chamber of the Parliament:

"Herewith I am sending you the promised note on reprocessing as well as the corresponding ECN-report.[161] The main conclusion of the ECN-report is that from the point of view of environmental impact and proliferation risks there is no preference for either reprocessing or direct disposal; however, from the point of view of costs and cost certainty there is a distinct preference for reprocessing.

Therefore, the government is of the opinion that there are no ponderous and urgent reasons to modify the Electricity-sector's current policy, which is based on reprocessing. I therefore intend - after consultation with the Chamber on the underlying note - to assent to GKN's[162] request to increase its current reprocessing contract with BNFL by 4.5 ton of uranium, as well as to the inclusion in that agreement of 76 irradiated MOX-fuel rods that are presently in storage at GKN. Thereby, the remaining (spent fuel) from the Dodewaard nuclear plant is covered by contract."

The diminishing interest for reprocessing in Europe is due to a number of factors. First of all, the expected development of fast breeder reactors requiring significant quantities of plutonium fuel did not materialise.

Secondly, the cost of reprocessing under the existing cost plus fee contracts has turned out to be at least twice as high (in real terms) as anticipated in the mid-seventies, while since then the prices of uranium supplies and enrichment services have drastically diminished.

And third, reprocessing plants have been the target of criticism by environmental pressure groups for their release of radioactive effluents in the sea and the atmosphere. The rising influence of Green parties in European politics have clearly influenced many governments not to support the reprocessing option. Notwithstanding the political realities, however, the technological reality is that, over the last ten years, the operating performances of existing reprocessing plants in La Hague and Sellafield have improved significantly. For example, the volume of high and medium level wastes, properly conditioned, has been diminished by a factor of 6, from 3 m^3 per ton of spent fuel initially to 1 m^3 in 1995 and 0.5 m^3 at present. By comparison, the volume of encapsulated spent fuel in the once-through cycle option is about 2 m^3 per ton of spent fuel.[163]

---

160 *Rapport sur l'Aval du Cycle Nucléaire*, Christian Bataille et Robert Galley, Députés, Office Parlementaire d'Evaluation des Choix Scientifiques et Technologiques, June 1998.

161 ECN: Netherlands Energy Research Foundation.

162 GKN is the operator of the Dodewaard nuclear power plant.

163 *Rapport sur l'aval du cycle nucléaire, op. cit.*, p. 63.

In 1997, the volume of conditioned low level waste (LLW) resulting from the operation of the La Hague reprocessing plant has been around 1.4 m^3 per ton of spent fuel. Therefore, even if one adds the volume of LLW, of intermediate/medium level waste (ILW) and of high level waste (HLW) - which is really adding apples and pears - the total volume of reprocessing wastes still does not exceed the volume of encapsulated spent fuel (either in the present German or Swedish concepts).

These facts are significantly different from the figures often mentioned by those opposing the reprocessing option who claim that "even today the total volume of conditioned reprocessing waste is about 20 m^3/t HM, while the volume of spent fuel in a German Pollux cask is about 2 m^3/t HM."[164] This indicates that the potential of technical progress is often underestimated. But what is more astonishing are statements such as: "However laudable these efforts (to reduce wastes' volumes), they still ignore the enduring problem of low active wastes which account for perhaps half of total reprocessing waste management and disposal costs."[165]

Such an assertion is in no way corroborated by the facts. Indeed, the cost of storage and disposal of 1 m^3 of conditioned LLW is at least 20 times less than the cost of storage and disposal of 1 m^3 of conditioned ILW and HLW (in proportion to their arising from reprocessing). Therefore, LLW management represents less (probably significantly less) than 15 % of the total waste storage and disposal cost, and is not an "enduring problem", since in the European Union alone 11 LLW disposal facilities are presently in operation.[166]

Also, the releases of radioactive effluents have been drastically reduced over the years, as can be seen in the table 1. Efforts to further reduce effluent discharges in the environment are permanent.

Year	**α-activity** (TBq/ton uranium)	**β-activity** excluding tritium (TBq/ton uranium)
1979	5.6 10^{-3}	9
1985	2.1 10^{-3}	3
1988	1.2 10^{-3}	1.4
1993	0.1 10^{-3}	0.08
1997	0.03 10^{-3}	0.024

Tab. 1. La Hague reprocessing site: liquid effluents released in the sea.

164 "Industrial and Environmental Rationales for Reprocessing", Frans Berkhout, SPRU, Global 1995, Versailles, September 11-14,1995, p. 414.

165 *Ibid.*, p. 414.

166 *Programme Indicatif Nucléaire*, European Commission, September 1997.

2.2.2 The proliferation risk

But the most important question concerning the reprocessing/recycling fuel cycle is whether it is, in the end, more or less vulnerable to the proliferation risk than the once-through fuel cycle. An analysis made on this subject by Professor André Jaumotte is worth quoting[167]:

"The reprocessing/recycling cycle is often considered to be more vulnerable to theft or diversion than the once-through cycle essentially because plutonium oxide is separated out and transported from the reprocessing plant to a MOX fuel fabrication facility.

It needs to be repeated, however, that transport of plutonium takes place under the most stringent international safeguards and physical protection measures. Should this not be considered safe enough, there exist a number of means to increase in the future the proliferation resistance of the reprocessing/recycling fuel cycle, for example:

co-processing, a process in which plutonium oxide recovered at the reprocessing plant remains at all times mixed with a much larger amount of depleted uranium oxide, thereby preventing its use without further processing even for making a crude nuclear weapon. Co-processing increases by 10 times or more the volume[168], that would have to be diverted undetected in order to reach a weapon's critical quantity of plutonium;

co-locating MOX fuel fabrication and reprocessing facilities, a simple means to avoid shipment of plutonium powders, thereby suppressing the risk of diversion during such transports;

reprocessing and MOX fuel fabrication could be restricted to a small number of countries such as, for instance, the five nuclear-weapons states and possibly Japan, provided they would be open to an international share-holding with a real right of control. All these plants would be subject to strict monitoring of regional and international safeguards, as is the case today in the European Union. The proliferation risk of such a regime would be virtually non-existent.

The comparison between the once-through and reprocessing cycles does not end here. It has been recognised by the American National Academy of Sciences that plutonium is least vulnerable to theft when diluted with uranium and highly radioactive constituents such as the fission products in spent fuel. This large self-protecting barrier is clearly effective for many decades after the spent fuel has been unloaded from the core of a reactor.

However, radioactivity naturally decreases with time. After about 200 years, the radioactivity of the spent fuel will have decreased below the 1Sv/h threshold that the US National Academy of Sciences has defined as "self protecting" from a proliferation point of view. This means that geologic repositories of spent fuel could become reasonably accessible plutonium mines and, therefore, a real concern to the international

---

167 "The non-proliferation benefits of reprocessing: swimming against the tide", Prof. A. Jaumotte, *Nuclear Europe Worldscan*, 7-8 1997.

168 Indeed, mixed oxide fuel used commercially in pressurised water reactors contains less than 10 % plutonium.

community. Radioactive wastes from reprocessing, on the other hand, are segregated into low-, medium- and high-level wastes containing only infinitesimal traces of plutonium. These wastes, after appropriate conditioning, can be stored and thereafter disposed of either in shallow land or in deep geologic formations without necessitating any further safeguarding.

The reprocessing/recycling option is, therefore, in the longer term, more proliferation-resistant than the once-through cycle."

2.3 All In My Back Yard (AIMBY) or Regional Centres?

Professor Jaumotte's analysis continues as follows:

"Under the pressure of public opinion, often unilaterally informed by vocal environmental fundamentalist groups, politicians all around the world, have accepted as self-evident that each country should dispose of its radioactive wastes within its national borders.

At first sight this AIMBY policy seems sound and responsible. But is it really? There are today more than 30 countries having to manage spent nuclear fuel, some with large and others with rather small quantities. Following the AIMBY policy, each of these countries will have to dispose of radioactive wastes at home.

Clearly if a disposal site is considered safe for indigenous radioactive wastes, it is also safe for identical foreign wastes. Therefore, it would appear more reasonable, in fact safer and less expensive, to have half a dozen large regional radioactive waste disposal centers rather than more than 30 national ones spread around the world.

Combining the No-Reprocessing and the AIMBY concepts would imply that there will be, spread in more than 30 countries, final disposal sites with encapsulated spent fuel containing significant amounts of plutonium that after some 200 years could be more or less easily retrievable. Is such a situation acceptable from a non-proliferation point of view? If not, what is the alternative?"

Professor Jaumotte then demonstrates that, by reprocessing only once all spent nuclear fuel produced in the world during 35 years and recycling the recovered plutonium as MOX fuel, one would end up with some 100 000 spent MOX fuel elements instead of 700 000 standard UO_2 spent fuel elements otherwise.

"To put this in perspective: 100 000 spent MOX elements represent about one-half the number of spent fuel elements that are expected to be produced in the US alone under the once-through cycle approach, and therefore one-half of the fuel elements which the US projects having the capacity to dispose of in its final repository.

This extreme example shows that, by having a very few well-safeguarded reprocessing/MOX facilities under regional and international control, and by disposing of a limited number of spent MOX fuel assemblies in the US and possibly one or two other places, one could avoid leaving to future generations a world where spent fuel containing plutonium would be stored in more than 30 countries."

The merits of regional or international spent fuel storage, encapsulation and disposal facilities have been recognised by a number of experts, even though such concepts are today usually not considered as politically correct.

In Canada, for instance, the idea of accepting foreign spent fuel for disposal on a commercial basis has been supported by some officials. The possibility was considered "of integrating power plant sales with waste management services to give the Canadian industry a unique advantage in the export market, and the possibility of importing spent fuel from countries that purchase Canadian uranium."[169]

In the US a similar suggestion is supported by W.K.H. Panofsky: "The proposal that nuclear reactors furnished by nuclear weapons states to potential proliferants be accompanied with an obligation to return the spent fuel to the originator should be an important move. This obligation is inherent in the proposal to furnish LWRs to North Korea and appears to apply to the proposal for Russia to furnish such reactors to Iran."[170]

In Russia, the Ministry of Atomic Energy (MINATOM) "would like to be able to change Russian law to permit it to offer reprocessing contracts to foreign utilities under which Russia would keep the waste and plutonium separated during reprocessing – meaning that utilities could send their spent fuel to Russia and never have to worry about it again."[171]

It was indeed confirmed in August 1998 [172] that a new law, which would permit the import of foreign spent fuel to Russia for the first time since 1992, looks likely to be adopted in the near future by the State Duma. The new legislation could result in Russia dramatically increasing its share of the world market for spent fuel management. Russia would use not only its technological experience and industrial capacities for spent fuel reprocessing but also its geographic advantages: large, poorly populated areas and the low seismic activity of most of its territory.

In Belgium, after an extended inquiry on waste disposal by a special Commission, the Senate recommended in 1991 that "although today that may seem utopian, a study should be launched at the European level on the conditions and means to reduce the number of high level waste geological disposal sites and on the localisation of such sites".[173]

Most recently, the Director General of the International Atomic Energy Agency (IAEA), Mohammed El Baradei, urged member states to focus greater attention

---

[169] *Nuclear Fuel Waste Management and Disposal Concept*, Canadian Environmental Assessment Agency, February 1998, p. 81.

[170] "Tension between nuclear proliferation danger, nuclear energy, and the disassembly of nuclear warheads", W.K.H. Panofsky, VIII International Amaldi Conference, Piacenza, Italy, October 5-7, 1995.

[171] "Reducing the threat of nuclear theft in the former Soviet Union", John Holdren, *Arms Control Today*, March 1996.

[172] NUCNET, 14 August 1998, *Business News*, n° 97.

[173] *Sénat, Commission d'information et d'enquête en matière de sécurité nucléaire*, Rapport final et recommandations, 12 July 1991.

on the issue of radioactive waste disposal and not rule out regional or international solutions.[174]

There is no doubt, that regionally or internationally managed spent fuel storage, encapsulation and disposal facilities would represent the best solution from an economic and, more importantly, from a non-proliferation point of view.

The United States, Russia and Euratom should devote more attention to these concepts and offer an acceptable solution for the spent fuel management of countries with modest nuclear power programs. As said by Professor Jaumotte: "To be or not to be politically correct is often only a question of time."

3. Economic Considerations

Concerning the economic aspect of spent fuel management, it is usually acknowledged that, calculated on a discounted basis, the once-through fuel cycle is the least costly option, mainly because it postpones the greatest part of the expenses in comparison with the reprocessing/recycling option. But the Nuclear Energy Agency (NEA) of the Organisation for Economic Co-operation and Development (OECD) concluded in a 1994 report[175] that "In light of the underlying cost uncertainties, this small cost difference between the reprocessing and direct disposal options is considered to be insignificant, and in any event, represents a negligible difference in overall generating cost terms."

It does not seem worth discussing this matter in detail here since, in the short to medium term, it is most probable that, at least in Europe, the market will decide. Indeed, in the present environment of oversupply in uranium, if reprocessing and recycling do not prove to be economically attractive, even for an industry that operates largely depreciated reprocessing plants and MOX-fuel fabrication facilities, then it is to be expected that, in a deregulated electricity market, the electrical utilities will opt in favor of spent fuel storage as being the cheapest way to keep their options open for the future.

Conclusions

The International Nuclear Energy Academy (INEA), whose members are internationally recognised for their experience and expertise in nuclear science, engineering and management, published in April 1997 a report reviewing the "Options for the Management of Spent Nuclear Fuel".[176] This report was intended to be an authoritative reference, recording the current positions on the technological and operational aspects of spent fuel management and on related issues such as non-proliferation and economics. The conclusions of the INEA report are the following:

---

[174] NUCNET, 21 September 1998, *Business News*, n 340.

[175] *The Economics of the Nuclear Fuel Cycle*, NEA/OECD, 1994.

[176] International Nuclear Energy Academy, Review Report, *Options for the Management of Spent Nuclear Fuel*, 3 April 1997.

- "Both reprocessing with recycle and direct disposal are viable options for the management of spent nuclear fuel and neither would present a bar to the further development of nuclear power; they can both meet the necessary safety and environmental standards.

- Interim storage of spent fuel is technically feasible even for the long term; it is a necessary precursor to either reprocessing/recycling or direct disposal and it maintains both options open.

- The reprocessing/recycling option is already in operation and direct disposal is likely to prove technically feasible - both options are likely to be pursued in parallel over the coming years and both can be operated without giving rise to an unacceptable environmental impact.

- Both reprocessing and direct disposal facilities are capable of being safeguarded and neither option presents a main proliferation risk; the main risk is likely to be clandestine plutonium or highly enriched uranium production outside international safeguards.

- So far as public acceptability is concerned, the main issue relating to spent fuel is the geological disposal of radioactive waste, and this is common to both the reprocessing/recycling and direct disposal options. Transport is also a concern, but as in waste disposal, the technology is well developed and the safety problem is mainly one of perception.

- The choice between the options will be dependent on many factors, and different countries and even utilities within a country may come to different conclusions according to the individual circumstances pertaining at the time.

- Economic differences between the two options could be outweighed by strategic/policy issues - the challenge with reprocessing is to reduce costs still further, the challenge with direct disposal is to reduce the uncertainty in costs.

- The world's supply of economically recoverable uranium could be exhausted within the foreseeable future unless the policy of recycling with fast reactors is available - this is an important factor in terms of long term energy resources."

The above conclusions are objective and well balanced and largely confirm our own findings. Based on the facts analysed in this chapter, and those in Chapter 13 dealing with the disposal of separated plutonium stocks, it is recommended that, whether or not new reprocessing commitments are delayed by European or Japanese utilities, excess stockpiles of civilian reactor-grade plutonium should be disposed of as MOX-fuel as rapidly as possible. Responsible national authorities should ensure neither undue political delays in licensing additional MOX-fuel fabrication capacity nor delays in allowing the use of MOX-fuel in an increased number of LWRs.

And, last but not least, more efforts should be made in Europe, the United States and elsewhere to develop regional or international spent fuel storage, encapsulation and disposal facilities.

Chapter 11 The Nuclear Fuel Cycle: Does Reprocessing Make Sense?

by
Richard L. Garwin*

Introduction

No more nuclear power plants are being built in the United States, and France (with 80 % of its electrical energy from nuclear) has an over-capacity of probably 25 % of its nuclear power plants. Let's look for an example of the near-term future of nuclear power elsewhere, where we have almost a clean state - China. China's plans for modernization and growth include a major expansion of nuclear power, which, even in the year 2050 is expected to provide only a small fraction of the electrical energy. China is fortunate in its size that allows economies of scale, in its substantial resources of technical and scientific personnel, and in its economy that has provided massive investments from abroad and a trade surplus that enables China to buy almost anything it needs from abroad, with a proper assignment of priorities. China's growth depends upon expenditures that contribute to growth in the most efficient fashion, and these efficiencies and economies (given always environmental acceptability) guide this discussion.

For the next 10 or 20 years, China's nuclear power will have to come from reactors of standard type, in operation the world over. This means for the most part light water reactors, for which I sketch the nuclear fuel cycle in the United States (which has more than 100 operating commercial power reactors) and in France, with about half that many.

1. The US Fuel Cycle

The fuel for all water reactors (both boiling-water reactors and pressurized-water reactors) is produced in a remarkably similar fashion. This is no surprise, because a market for fuel materials and fuel elements must be governed by standards, and in the United States these are the so-called ASTM standards (American Society for Testing Materials). Since low-enriched uranium fuel is a commodity on the world market, it is essential to have standards that are internationally harmonized. The sequence of steps for the fuel cycle is:

---

* Richard Garwin is Philip D. Reed Senior Fellow for Science and Technology at the Council on Foreign Relations, a Fellow Emeritus of the IBM Thomas J. Watson Research Center, and Adjunct Professor of Physics at Columbia University.

1. Mining and milling of uranium ore.

2. Beneficiation, or purification, resulting in crude uranium oxide U_3O_8, also called "yellow cake".

3. Conversion to uranium hexafluoride (UF_6). This is a solid at room temperature, but a gas above 57°C, and is particularly suitable for uranium enrichment, because fluorine in nature has only a single isotope. So the difference in mass of the various UF_6 molecules in the gas is due entirely to the difference in mass between the molecules containing ^{238}U and the 0.7 % of molecules (in normal uranium) which contain ^{235}U.

4. Isotope enrichment. This is done in the United States by gaseous diffusion in which the UF_6 molecules penetrate through a porous barrier at slightly different rates (because of the 3-unit mass difference out of almost 500 atomic mass units, the ^{235}U molecules penetrate through the barrier about 0.3 % faster than the molecules containing ^{238}U). The major alternative in uranium enrichment is the use of a gas centrifuge. The modern high-performance gas-centrifuge isotope enrichment plant consists of many tall cylinders each spinning in a vacuum at some 1500 m/s surface speed; the rotation produces high equivalent gravitational fields, and separation factors per stage are larger than in the gaseous diffusion plant. Some of the European enrichment plants and all those in Russia now use the centrifuge process.

5. Conversion to oxide. The UF_6, which is chemically reactive with air or water, is converted to UO_2, a black powder of grain size appropriate for pressing into ceramic pellets. These pellets are then baked at high temperature (so that the fine particles "sinter" or form strong bonds through surface diffusion), are then ground to size, and are ready to be fabricated into fuel rods.

6. Fuel fabrication is completed by loading the ground pellets into thin-walled zirconium alloy tubes, to constitute the fuel rods some 5 m long and about the diameter of a pencil. Several hundred such rods are carefully mounted with spacers in a fuel assembly, for insertion in the reactor.

7. Burning in the reactor. A typical fuel assembly is inserted into the reactor and operated at essentially full power for four years. Actually, every 12 months the reactor is shut down for a month, so that one-fourth of the core can be replaced with fresh fuel. Other elements may be "shuffled" in order to equalize the neutron exposure and the fissions obtained from each gram of fuel. The fuel element is typically burned to 40 000 megawatt days per ton of heavy metal[177] and removed from the reactor. Some reactors are operated for 18 months, with one-third of the fuel replaced during the shutdown.

---

[177] Megawatt days per ton of heavy metal: MWd/tHM.

8. The 20 tons removed each year, of the typical 100 tons of fuel in the reactor, is transferred, under water, to the "swimming pool" storage at the reactor, where it is immersed together with other fuel elements. The fission-product decay heat is such that the fuel element would glow red-hot if it were removed from the water, in which it is cooled by natural circulation. Within the reactor, of course, the fission heat is removed by flowing water coolant, which either converts to steam at the top of the reactor pressure vessel in a boiling water reactor, or in a pressurized-water reactor is transferred at high pressure in the primary loop to the steam generator, where it passes on the heat to the secondary loop, where water is converted to steam for operating the turbine. The turbine and the ensuing electricity are non-nuclear aspects of the plant, and are not part of the fuel cycle.

9. Reactors have generally been built with large pools adjacent to the reactor so that fuel elements removed from the reactor after four years of fission and heat production can be stored safely without overheating. After a couple of years of cooling in the "at-reactor pool", the fuel elements are loaded into a cooled cask for transport to an intermediate or permanent storage; in reality, almost all US spent fuel is still in the storage pools at the reactors. Typically, the fuel will be stored for 10 or 20 years or more in above-ground storage before it is disposed of. In the United States, the storage is to take place in a heavy steel cask - about 24 fuel elements to a cask - which serves for safe transport of the fuel and for dry-cask storage, so that the fission-product decay heat (about 1 kW per fuel element at the time the elements are loaded into the cask) can diminish over the years. The amount of fission products that can be put into the repository is determined primarily by the gradual heating of the rock, so the cost is reduced by allowing much of the fission product radioactivity (hence, heat) to decay in dry cask storage, before emplacement in the repository.

10. Eventually, the plan is to place the fuel, within its storage cask, in a mined geologic repository. The site that the US has selected for its repository for the disposal of 80 000 tons of spent power reactor fuel and other high level waste (HLW) is Yucca Mountain in Nevada. The US disposal program has been fraught with political and technical difficulties, and the Yucca Mountain repository will not open until the year 2010 at the earliest. Reactor operators have had to modify their storage pools to hold more fuel elements, and they complain that they are running out of storage capacity.

Every one of these steps contributes to the cost of the fuel cycle, and is regulated, in order to maintain an acceptable level of protection against hazard to the public and to the workers. There is also associated with each of the steps some residual radiation exposure to the workers and also radiation exposure to the public.

2. The French Nuclear Fuel Cycle

It is evident that much of this fuel cycle is identical with that used in French reactors, where the fuel, after being removed from the reactor and allowed to cool in the at-reactor swimming pool, is then transported to intermediate storage at the reprocessing plant. There the fuel elements are disassembled by machine, the rods chopped automatically (all under water) and the "hulls" (the sheaths and the ends of the rods and the other fuel element hardware) are cleaned and collected for disposal. The uranium-oxide ceramic fuel pellets are then dissolved in strong chemical reagents, with a view to purifying to an extreme degree the uranium and the plutonium from the fission products; less than one part in 10 million of the fission products remain in the uranium or plutonium product, and 99.9 % of the plutonium is removed from the fission product waste. The uranium oxide, typically now 0.9 % ^{235}U (compared with 0.7 % ^{235}U in natural uranium[178], or 4.4 % ^{235}U in the low-enriched uranium (LEU) fuel as it was loaded into the reactor) is sent from France to Russia for enrichment. The enriched LEU is returned to France for fabrication into fuel for the reactors. France does not wish to contaminate its large-scale gaseous diffusion plants with traces of plutonium or with uranium for recycling.

The plant at La Hague, operated by COGEMA (*Compagnie Générale des Matières Nucléaires*, in charge of the fuel cycle in France), processes 1600 tons of spent fuel per year in two large modules of 800 tons/year each. The spent fuel from low-enriched uranium loaded into the reactor contains about 1 % plutonium, so about 200 kg to 250 kg from a 20 to 25-ton annual download from each reactor. La Hague thus separates about 16 tons of so-called reactor grade plutonium (R-Pu) per year. This is produced as plutonium oxide (a dry powder) which is then welded into small steel cylindrical containers each holding about 2 kg of plutonium.

Compared with weapon-grade plutonium (W-Pu), which is typically 94 % ^{239}Pu and no more than 7 % ^{240}Pu, reactor-grade plutonium is some 60 % ^{239}Pu, 25 % ^{240}Pu, 9 % ^{241}Pu, 5 % ^{242}Pu, and 1 % ^{238}Pu. Only the odd isotopes 239 and 241 (i.e., those with an odd number of nucleons) of plutonium are fissile - that is, are subject to fission by slow neutrons. Five of these small containers are sealed into an outer steel cylinder for protection and storage. In this form they are stored and eventually transported to the fuel fabrication plant, where the plutonium oxide is mixed with uranium oxide, and fabricated into so-called "Mixed Oxide" (MOX) pellets.

But the fabrication process for MOX fuel is considerably more hazardous or costly than that for uranium fuel. The half-life of ^{239}Pu is 24 000 years, in comparison with 4.5 billion years for ^{238}U and 0.7 billion years for ^{235}U. The Environmental Protection Agency regulations mandate that for reactor uranium there should be less than one gram of material in 38 million cubic meters of air (about 50 000 tons of air). The

---

178 Because of the presence of ^{236}U in this uranium from reprocessing (produced by parasitic capture of neutrons in ^{235}U that did not lead to fission), this uranium containing 0.9 % ^{235}U is less valuable than natural uranium; the high avidity of ^{236}U for thermal neutrons with no possibility of fission requires larger ^{235}U concentration in fuel made from recycle uranium than from natural uranium.

corresponding limit for reactor plutonium is one gram of plutonium in 2.9 10^{13} cubic meters of air - a million times more stringent. The resultant protective measures in a MOX plant lead to a fabrication cost of some $2000 per kilogram of fuel -"$2000/kgHM" where "kgHM" is read "kilograms of heavy metal" - in order to provide a measure applicable both to MOX and to low-enriched uranium.

To furnish a fissile content for the MOX fuel comparable with that in normal uranium, about 5 kg of spent uranium fuel must be reprocessed to yield enough plutonium for 1 kg of MOX. If we estimate $1000/kg of spent fuel for reprocessing, this results in a reprocessing component of MOX fuel cost of $5000/kgHM, for a total MOX fuel cost, including fabrication, of about $7000/kgHM in comparison with current purchase price of uranium fuel ready to load into a reactor of some $1400/kgHM.

Let's take up in some detail the relative merits of reprocessing and recycle of MOX fuel, in comparison with the once through "direct disposal" approach to which the US nuclear power industry is committed.

3. Criteria For Choice

China's size and wealth of technical resources means that it can consider undertaking developments that would be unthinkable for a country the size of Sweden or Switzerland. But just because of the large number of reactors that are to be built, the choice of an option of higher cost or greater hazard would constitute a greater total burden. China can afford to import technology and capital to jump-start its nuclear program. In this regard, the French approach is a useful model. France built the early units of its modern pressurized-water reactors (PWRs) under contract with Westinghouse, and soon developed an indigenous capacity to build reactors in France and all over the world.

As will be seen from the above analysis, I do not believe that France should have undertaken reprocessing of this Light Water Reactor (LWR) fuel - an approach that caused (negligibly) greater radiation exposure and imposed (substantially) higher costs on the reactor operators and consumers of electrical energy in France.

Many choices need to be made in the Chinese nuclear power program, but some of them are more global than others. We limit our consideration to LWRs and initially to the choice between:

1. direct disposal of LWR spent fuel into a mined geologic repository, or

2. reprocessing, with recycling of plutonium and disposal of vitrified fission-product waste.

The criteria to be used are radiation and accident hazard, the problem of proliferation of nuclear weaponry, and economics. I will dispose of the first of these two in the special case of China with a few words.

3.1 Radiation and accident hazard

In the 1975 "Report to the American Physical Society on the Safety of Light Water Reactors", my colleagues and I looked at the probability of catastrophic accidents. And in our 1977 book, "*Nuclear Power Issues and Choices*", Panofsky, Keeny, and I (and others) summarized nuclear power as being safer than coal, taking into account our best estimates of the likelihood of catastrophic accidents, assuming that nuclear power plants were operated according to a standard that we perceived was being followed in the United States. The 1979 core meltdown at Three Mile Island was in accord with our estimate of probabilities, although it was a shock to many who had maintained that meltdown was essentially impossible. Adding the reprocessing industry to the nuclear power operation clearly adds to the probability of an accident, but not, in my judgment, to the extent that it overwhelms the economic calculation.

3.2 Proliferation potential

In the case of China - a nuclear weapons state - the separated plutonium that would be involved in a large-scale nuclear power industry would greatly exceed that which has been produced for nuclear weapons. This would impose a cost of safeguarding and security to prevent the theft of such material and its transfer abroad. But there is no doubt that the reactor-grade plutonium obtained from reprocessing LWR spent fuel can readily be used to make high-performance, high-reliability nuclear weaponry, as explained in the 1994 Committee on International Security and Arms Control (CISAC) publication.[179] While R-Pu cannot be directly substituted for the same metal nuclear weapon part made of weapon plutonium, the problems of making nuclear weapons from R-Pu are little different from those encountered in making a nuclear weapon from W-Pu.

A principal problem of the reprocessing approach is that the commercial firms providing or using reprocessing permit or require spokesmen to assert that R-Pu cannot be used to make nuclear weapons, thus encouraging the separation and transport of plutonium elsewhere without adequate protection against theft. I assume that China would provide adequate security and safeguards for its domestic reprocessing, so we go on to the economic comparison.

3.3 Economics

The economics of nuclear power are dominated by capital and interest charges on the reactor, together with carrying charges on this investment. A good treatment is found

---

[179] *Management and Disposition of Excess Weapons Plutonium*, Report of the National Academy of Sciences (NAS), Committee on International Security and Arms Control (CISAC), W.K.H. Panofsky (Chair), 1994, pp. 32-33.

in the 1995 CISAC volume.[180] Using a nominal plant lifetime of 30 years, and a real cost of money of 7 % per year, and assuming that a modern power reactor can be built for some $3.4 billion:

1. We find a reactor-cost component of electrical energy of $0.0386 per kWh.

2. The non-fuel operations and maintenance costs for such an operation are of the order of $0.0140/kWh.

3. The fuel cost for 4.4 % ^{235}U is given as $0.0052/kWh.[181]

The total cost of electrical energy at the "busbar" is the sum of these three components: $0.0578/kWh - about 6 US cents per kWh. The CISAC 1995 report shows an uncertainty in similar estimates of about ±10 %.[182] The costs of transmission and distribution (T&D) of the electrical energy add considerably to this cost, and if the reactor does not operate at the assumed capacity factor, costs can be much higher in view of the large ratio of capital cost to fuel cost. Note that the situation is quite different for a fossil-fuel plant, for which the fuel cost greatly exceeds the capital-cost charges. Comparison of the fuel-cycle costs between direct disposal and reprocessing-and-recycle allows us to estimate the baseline extra cost incurred with reprocessing.

In the reprocessing of spent fuel and recycling of plutonium as MOX ("R&R"), some 5 kg of spent LEU fuel must be reprocessed to yield the plutonium that can provide 1 kg of MOX equivalent to 1 kg of fresh LEU fuel. According to the CISAC 1995 report, reprocessing costs about $1000/kgHM.[183] Fuel fabrication of MOX costs about $2000/kgHM. Thus, about 20 % of the fuel is replaced by MOX obtained at a cost of about $7,000/kgHM rather the $1400/kgHM that would be paid for LEU fuel of similar performance. The extra cost of $5600/kgHM may be considered an increase in average fuel cost of $5600/5 or about $1120/kgHM for all the fuel - an increase of 79 % in fuel cost that then contributes an additional $0.0041 per kWh to the cost of electrical energy. The choice of reprocessing and recycling would thus increase the busbar cost by about 7 %.

This additional cost would be acceptable (it is about 10 % as big as the capital-cost component due to the reactor), if it brought real benefits. But it does not. The strongest argument advanced for reprocessing and recycle is that the supply of uranium is finite and that the 20 % saving in raw uranium will extend the supply. We need now to address the fuel supply question and the timing of reprocessing. We will see that relating the MOX cost to a percentage increase in the cost of "all fuel" leads to a substantial

---

180 *Management and Disposition of Excess Weapons Plutonium: Reactor-Related Options*, Report of the NAS, CISAC, Panel on Reactor-Related Options for Disposition of Excess Weapons Plutonium, 1995, see ch. 6, especially pp. 303-324.

181 See table 6-12 of *Management and Disposition of Excess Weapons Plutonium: Reactor-Related Options*, *op. cit.*

182 *Management and Disposition of Excess Weapons Plutonium: Reactor-Related Options*, *op. cit.*, p. 310.

183 "HM" refers to "heavy metal", used in order to have a term that includes both Pu and U (and plutonium oxide and uranium oxide).

policy error. We will instead calculate the "permissive cost" of enriched uranium to displace that MOX fuel.

4. Nuclear Fuel Supply

Let's now consider the future energy requirements of China. Table 1 shows the projected use of electrical energy by China in 1995 and 2050.

Of course, 55 years is a long time for undisturbed growth, but it is notable that these projections will still leave nuclear power providing only 13 % of China's electricity in 2050.

Each LWR fissions about one ton of heavy metal per year (in the 20 tons of fuel that it is fed per year) and produces a ton of fission products. Some 200 tons of natural uranium is needed to supply the low-enriched uranium fuel for a reactor year of operation. Thus, the 240 reactors China expects to have in 2050 would consume about 48 000 tons of natural uranium per year, of a world uranium reserve ("proven and very likely") that is estimated as some 3 million tons. Of course, China's reactors are not alone - the world inventory now being some 300 such full-size power reactors. Clearly, a world population of 2000 reactors would consume 400 000 tons of natural uranium per year - and 3 million tons in a mere 8 years.

Year	**Population**	**Energy Used**	**Electricity Production**				
			Total	***Hydro***	***Fossil***	***Nuclear***	***Renew-able***
	(billions)	(GTCE/y)[184]	(GW(e))				
1995	1.2	1.24	200	49	149	2.1	0
2050	1.5 - 1.8	4.0	1600	250	1100	240	100
Annual Growth	0.52 %	2.13 %	3.78 %	2.96 %	5.66 %	8.62 %	-

Tab. 1. The use of electrical energy by China: 1995 and 2050.
Source: CISAC Beijing discussions, 1997.

But the conclusion is certainly not that one should implement reprocessing and recycling of plutonium in light-water reactors, which would reduce uranium consumption by only 20 %, at considerable cost. We have seen that this 20 % reduction of demand cost of 79 % corresponds to an increase of cost of the fuel displaced by $7000-

[184] 1 TCE/y (ton coal equivalent per year) is about equal to 1 MW(th) (megawatt thermal energy). See appendix of this book for details.

\$1400 = \$5600/kgHM. This increase of \$5600/kgHM of fabricated fuel could be produced alternatively by paying a higher price for that particular natural uranium, while the cost of conversion, enrichment, and LEU fuel fabrication remain the same. Since natural uranium contains 0.71 % ^{235}U, and the "tails" of the enrichment plants contain about 0.2 %, about 8.23 kg of natural uranium is required to produce 1 kg of LEU; the \$5600/kgHM increased cost of that substitute LEU fuel would pay \$680/kg of natural uranium additional to the current cost of some \$10 to \$20 per kg of uranium.

This factor 34 or more increase in price for any raw material in general produces a great increase in supply. Commercial firms have no interest in exploring for such high-cost resources, so our understanding of the amounts of uranium available at these higher prices is not firm. However, John Holdren is among those who have estimated the magnitude of this important energy resource.[185] Instead of 4.5 million tons of uranium often cited as the "reserve", the exploitable resource is likely to be more in the range of 100 to 250 million tons of uranium at prices below \$350 per kg of uranium. A 1 GWe nuclear power plant requires about 1 ton of ^{235}U per year, from about 200 tons of natural uranium. A world with 2000 nuclear reactors instead of the present 432 would need 0.4 million tons of uranium per year; so a world resource of 100 million tons would serve such a population of light water reactors for 250 years. It is especially important to consider uranium from seawater, where there is 4 billion tons; two billion tons of this could be extracted without significant change of cost due to depletion.[186] We have noted that a Japanese program on the extraction of uranium from sea water results in a 1998 estimate of \$100 per kg, which would ensure the availability of inexpensive nuclear fuel for thousands of years.[187] The seawater uranium resource would supply a population of 2000 light-water reactors for 5000 years and breeders for 500 000 years.[188]

---

[185] John P. Holdren and R.K. Pachauri, *Energy*, ICSU, An Agenda of Science for Environment and Development into the 21st Century, Cambridge University Press, 1992, pp. 103-18.

[186] The measured 3 microgram/kg for an ocean 5 km deep and 500 million square kilometers (times 80 %) gives 4 000 million tons of natural uranium (NU), compared with assured resources at current prices of some 3 million tons of terrestrial NU. Assume 2 000 million tons of NU from seawater are available without a big increase in cost over the first 100 million tons. A reactor-year (1 ton of ^{235}U) requires 200 tons of NU, so 2 000 million tons is 10 million reactor-years for LWRs.

[187] In France, Foos and his team have published estimates of \$40-50/kg by the use of nanofiltration and ion exchange plastics. Tadao Seguchi, Director of Material Development at the Japan Atomic Energy Research Institute (JAERI) has reported million-fold concentration of seawater uranium in plastic adsorbents, and estimated a cost of recovery of seawater uranium of some \$100 per kg uranium (personal communication, May 23, 1998, reconfirmed by e-mail, October 23, 1998). T. Seguchi emphasizes that system studies have not yet been done, so this estimate is quite uncertain.

[188] The work of the JAERI group on extraction of uranium from seawater (Tadao Seguchi, 1998) begins to answer many questions about this resource. Actual trials of a plastic polyethylene-based absorbent for 20 days in seawater show concentration factors of a million for uranium, 2 million for titanium, 3 million for vanadium, and 60 million for cobalt. Each kg of adsorbent at the end of 20 days contains 3g U, 2g Ti, 6g V, and 6g Co. What is the permissive cost for recovering the adsorbent, flushing the uranium and restoring the adsorbent, for a resulting natural uranium cost of \$100/kg (assuming that the other recovered metals have no value)? Some 160 kg of adsorbent must be processed to retrieve 1 kg of U, so the processing cost per kg of adsorbent must not exceed \$0.60. There are other components of cost as well. For instance, 160 kg of adsorbent provide 1 kg of U in 20 days (or 1 kg of adsorbent provides 1 kg of U in 3200 days). If the life of the adsorbent is 10 years, its cost must not exceed \$100/kg. Since the cost of this

We draw two conclusions from this comparison between the costs of reprocessing and recycling on the one hand, and the "permissive cost" to increase supplies of uranium on the other hand:

1. It is very likely that increased supplies of uranium will be available for fuel cost increases far below those associated with reprocessing.

2. It is urgent for governments to do the exploration, not to exploit the uranium resource but to sample and to do enough development to obtain a better estimate of the supply vs. cost of uranium.

3. The uranium in seawater, equally available to all, is an extraordinary resource. It is urgent for governments to do the chemistry, chemical engineering, and systems studies to understand the cost of uranium from seawater, if we are to avoid very large investments in reprocessing that are likely to be unnecessary for hundreds of years.

4. So long as the capital cost of breeders exceeds that of light-water reactors, seawater uranium is likely to make breeders un-economic, since the fuel component cost of nuclear energy is so small in comparison with capital costs.

5. Contrary to the case with oil or gas fuel, both the low fuel cost for nuclear energy and the chemically inert nature of natural uranium or LEU make it practical and economic to stockpile supplies for many years. Concern about uranium supply in the mid-term may be eased by purchasing LEU from Russia, blended to 4.4 % ^{235}U (or even to 19.9 % ^{235}U) from the high-enriched uranium now excess to the Russian nuclear weapons. In this regard, it would be desirable to relax the rather arbitrary standard for ^{234}U content (set in the United States as 1.00 % of the ^{235}U). Increasing this limit by 50 % would allow blending of HEU directly with normal uranium, rather than requiring the production of 1.5 % LEU to blend with the HEU in order to stay within this limit.

Finally, we should recognize that modern LWR spent fuel will survive un-degraded for centuries in the planned storage casks, either in aboveground storage or in mined geologic repositories. It is a fundamental principle of economics that one should not spend money or resources earlier than necessary; in this case, reprocessing could be phased in 50 or 100 years from now if the cost of uranium rises as a result of land-based resources being less than the estimate, and if it turns out that uranium from seawater proves to be more difficult than now seems likely.

As we have seen, reprocessing spent fuel to obtain plutonium for recycle in light-water reactors is an extremely high-cost approach to extending the uranium fuel resource. If we assume that the purpose of reprocessing and recycling ("R&R") is to reduce uranium consumption, we see that we have paid $5600 to save the natural uranium that

---

modified polyethylene is more likely to be $5/kg than $100/kg, capital cost does not seem to be a large part of the total.

would otherwise have been used to make 1 kg of 4.4 % LEU fuel - some 8.23 kg of natural uranium. R&R thus costs $5600/8.23 = $680 per kg of uranium saved. Since uranium costs about $20-$40 per kg, this is an enormous cost, not justified until the cost of extraction of uranium rises to $700 per kg.

4.1 When the breeder reactor?

The economics of the breeder reactor can also be usefully explored in terms of "permissive cost". Consider a normal metal-cooled uranium-plutonium cycle breeder reactor with capital cost F times that of a normal LWR power plant. Begin with the elements of cost per kWh for an LWR:

$0.0386 /kWh	Reactor-cost component
$0.0140 /kWh	Non-fuel operations and maintenance costs
$0.0052 /kWh	Fuel costs for 4.4% LEU
______________+	
$0.0578 /kWh	Total costs: about 6 cents per kWh

Assume that F = 2. That is, a breeder power plant costs twice as much as an LWR, and suppose that the fuel cycle cost is zero - reprocessing, fabrication, etc. If non-fuel operations and maintenance are the same for the breeder as for the LWR, the energy costs will be:

$0.0772 /kWh	Reactor-cost component
$0.0140 /kWh	Non-fuel operations and maintenance
$0.0000 /kWh	Assumed fuel costs
______________+	
$0.0912 /kWh	Total costs: about 9 cents per kWh

When would such a breeder be competitive with the LWR? Only when the cost of uranium fuel for the LWR is much higher than now. In fact, the fuel cost would need to rise by $0.0334 / kWh for the breeder to have the same energy cost (on the assumption that F = 2). Recall that the $0.0052 /kWh fuel cost for the LWR was for LEU fuel cost of $1400/kg, so that it could rise by a factor of 6.4 to $9000/kg of LEU and still equal the breeder cost for energy. Since each kg of LEU requires some 8.23 kg of natural uranium (NU), NU costs would need to rise by $9000/8.23 = $1093/kg NU to make the breeder competitive. But we don't know that the breeder will actually be twice as costly as the LWR plant (F = 2). If we assume uranium price to rise by $225/kg from the present $20-$40/kg, the same analysis shows that the breeder power plant would need to cost less than 20 % more than the LWR power plant in order to be competitive. And if the breeder fuel cycle cost is more than zero, the capital cost of the breeder would need to be even less than this 20 % excess to compete with LWRs fed with LEU from natural uranium costing \$250 per kg. Also, the performance of the breeder (capacity factor) must be no worse than for the LWR.

This is not at all a negative view on the breeder, but it does say that breeders are not interesting until they cost no more than LWRs.

5. Disposal Of Spent Fuel Or High-Level Nuclear Wastes

As part of the nuclear fuel cycle, whether nuclear reactors continue to operate or not, the highly radioactive spent fuel will need to be kept from the biosphere for millenia and preferably for millions of years. Because of the fission-product decay heat, fuel elements removed from the reactor are universally kept under water in pool storage at the reactor. After some years, they may be shipped to intermediate storage or for reprocessing, which purifies the plutonium and uranium by a factor 10 million or 100 million from the fission products, and holds the solution of fission products for eventual vitrification. In this latter process, as practiced at La Hague and elsewhere, a stream of highly radioactive fission product solution is dripped onto powdered glass in a high temperature smelter, so that the fission products end up encapsulated in borosilicate glass melted into a stainless steel container that is then welded shut and cleaned. The resulting sheathed glass "logs" are then stored for some years and then deposited in a mined geologic repository.

Neither spent fuel in its disposal/storage cask nor glass logs have yet entered a mined geologic repository. That in the United States, at Yucca Mountain (Nevada) is delayed by lawsuits which prevent the Department of Energy (DOE) from disposing of the spent fuel from commercial reactors, for which it has been charging the electric utilities \$0.001 per kWh for ultimate disposal of the fuel. In France, the reprocessed plutonium and uranium, and the glass logs containing the fission products, are stored temporarily at La Hague and eventually all returned to the owner of the spent fuel that has been reprocessed. Vitrified fission products from reactors in France are stored at La Hague in glass logs; it is regarded as significant that these not be called "waste", and they are destined eventually for "underground laboratories" and not "waste disposal sites". But as yet, none have been so emplaced.

It is clear that competitive, commercially mined geologic repositories are needed for disposition of spent fuel in a "direct disposal" approach; they are equally needed for disposition of vitrified fission products in an "R&R" approach. In the US, there is Yucca Mountain; in France, there are several "underground laboratories". No country currently accepts other country's spent fuel or vitrified fission products for disposal. Thus, countries such as Switzerland and Sweden must do research and development (R&D) on their own projects. This situation is highly uneconomic and, clearly, a market is needed here. The solution is: competitive, commercial mined geologic repositories under International Atomic Energy Agency (IAEA) regulation and international protection to accept, for a fee, IAEA-approved spent fuel or vitrified fission products. China should offer this service and make money. The US would soon follow, as would Russia and many other countries. This would be very much in the interest of the nuclear reactor operator and the consumer; it would probably not be welcomed by those who profit from the present inflexibility of the industry.

Conclusion

Reprocessing of spent light-water reactor fuel and recycling of plutonium cannot be justified on the basis of reduced uranium needs, since they provide fuel at a cost comparable with low-enriched uranium from natural uranium costing $700 per kg in comparison with the $20-$40/kg now being paid. A breeder reactor that costs 20 % more than a light-water reactor power plant will be economic only when the raw uranium cost rises by more than $225/kg above present prices. Uranium from seawater would expand the resource by a factor of 1000 over the present proven reserves, and by a factor of 100 000 if breeders are eventually deployed. It is urgent for nations to do the technical work to understand the cost of acquiring uranium from seawater (estimated as some $100/kg U), or from terrestrial resources of lower grade and much higher cost than the present $20-$40/kg U. Industry has no present incentive to do such work, but it is essential to enable sensible national choices and to avoid what seem to be much more costly options such as plutonium recycle and breeder reactors. It would greatly help the consumer of nuclear electricity to have competitive, commercial, mined geologic repositories that would handle spent fuel elements and vitrified fission products, in a form certified by the IAEA.

Chapter 12 Why Reprocess? - A UK Case Study

by
Jack Harris*

Introduction

The question arises whether or not a verifiably-safe nuclear-weapon-free world is achievable while there continues to exist a global network of civil nuclear power stations. It is undeniable that the existence of a national civil nuclear power programme makes it vastly more difficult to ensure by inspection that any particular country is not also clandestinely manufacturing nuclear weapons. Nearly all the facilities for producing the vital fissile components of nuclear warheads can, apparently legitimately, be in place and operating - reactors to generate plutonium, reprocessing plant to separate this plutonium and at the other end of the fuel cycle, enrichment facilities, ostensibly in place to bring enrichment levels to those required by modern reactors. Should those who seek the complete removal of nuclear weapons from the world's arsenals also demand, albeit regretfully, the end of nuclear power? Before addressing this question in more detail it is helpful to review the chequered history of nuclear electricity, drawing particularly on UK experience.

1. Early Hopes for Nuclear Power

In the halcyon days of nuclear power, say the decade immediately following the first Geneva Conference on Peaceful Uses of Atomic Energy in 1955, individuals and groups who became involved in constructing nuclear power stations consoled themselves with four comfortable thoughts:

(a) the first five nuclear weapon states (US, USSR, UK, France and China, in chronological order) all started along the nuclear path by developing nuclear weapons; only subsequently did these efforts evolve into programmes for the construction of civil nuclear power stations (i.e. there were no early examples of nuclear weapons being produced from the products of "peaceful" nuclear power programmes);

(b) the plutonium produced in high-burn-up civil reactors was totally unsuitable for the production of nuclear bombs;

(c) the operation of a nuclear reactor could not result in a nuclear explosion, and

(d) as fossilised fuel became exhausted, the cost of uranium too would rise such that the fast reactor would become economically viable and this would justify the

* Jack Harris is Editor of Interdisciplinary Science Reviews and formerly Senior Section Head, Berkeley Nuclear Laboratories, UK.

formation of a reprocessing industry and in effect solve the world's energy crisis for at least a millennium.

2. Disillusionment

In fact all four of these comforting thoughts and aspirations turned out to be illusory or without substance:

(a) in 1974 India stunned the world by exploding a "peaceful" nuclear device which had been manufactured from plutonium generated in a Canadian Deuterium (CANDU) reactor supplied for peaceful purposes by Canada, America having provided the heavy water;

(b) in the 1970s it gradually became clear that plutonium of almost any isotopic composition was capable of sustaining a chain reaction and hence producing a nuclear explosion, indeed it was revealed that in the late 1960s the Americans had produced such an explosion with plutonium of roughly "civil" composition;

(c) the catastrophic eruption at Chernobyl was a *nuclear* explosion (hence its ability to contaminate on a global scale), and, finally

(d) the anticipated rise in fossilised fuel and uranium prices did not materialise so that even today the fast reactor remains hopelessly uneconomic and will remain so at least for the next several decades.

Even more importantly, particularly in those countries with highly developed democratic processes, such as parts of Western Europe (notably Scandinavia, but excluding France) and the US, there developed a widespread public fear and dislike of nuclear power and a virtual moratorium on the construction of new stations (this aversion, though strengthened by the accident at Chernobyl, had developed considerable strength by the mid-seventies, i.e. well before even the accident at Three Mile Island). In some countries, the antipathy towards nuclear power has led to the premature close-down of nuclear stations, most notably recently in Sweden and Germany. More generally, certainly as far as Western Europe and the US are concerned, if no new nuclear reactors are allowed to come on line then inevitably the industry will decline and disappear as existing stations sequentially are closed down on reaching their design lifetimes.

3. Nuclear Power Retains its Attractions in some Countries

It is, though, decidedly not the case that all that has to be done is to wait upon events and that, as part of a natural process, nuclear power will decline towards zero in tune with the anticipated reduction in nuclear arsenals. In the first place, there are countries where nuclear power is burgeoning or is already a vital part of the national economy, notably France and Japan and other Pacific Rim countries; and China has an ambitious future reactor building programme. World-wide, there are at present operating some 430 nuclear reactors supplying about 17 % of all electricity demands. It is inconceivable that all these countries would willingly abandon their nuclear reactors as part of a disarmament agreement. No doubt they would argue that sufficient safeguards were already in place to ensure that during no part of their fuel cycle was there any chance of fissile material being diverted for illicit purposes.

There are other factors which might well lead to a revival of the fortunes of the nuclear power industry, notably the concern over global warming and the contribution to it of carbon dioxide released by burning fossil fuel (see Chapter 2, by Fetter). The agreements reached at Kyoto require in many cases significant reductions in carbon dioxide emissions. In addition to such environmental concerns, it is quite possible that within a few decades the much heralded shortage of fossilised fuel could become a reality, with an associated rise in fuel costs and greater emphasis being placed on nuclear generation. It should be remembered that France and Japan have invested very heavily in nuclear power, not for environmental reasons, but as a reaction to their individual country's relatively low natural reserves of fossil fuels. Decades of uranium ore requirements can be purchased and stored relatively easily, so that the nuclear route is one which creates independence of outside suppliers.

In the case of Japan, the world's second largest economy, the dread of running out of energy amounts almost to paranoia; it should be remembered that it was fear of having fuel supplies cut off (at that time mainly oil) by American embargoes which was an initiating factor for their entry into the Second World War.

4. Tighter Controls needed on the Spread of Nuclear Power

It appears that advocates of nuclear disarmament must reconcile themselves to the continued existence of numerous nuclear power stations in many countries and must adjust their monitoring and verification procedures accordingly. There are ways, however, of minimising the difficulties the presence of nuclear generating stations can cause. Firstly, and most obviously, it may be necessary to renege on the undertakings implied in the 1955 Atoms for Peace Conference (and incorporated to a degree in the Non-proliferation Treaty, NPT) to share enthusiastically the benefits of peacefully-applied nuclear energy by, for example, assisting Third World countries to construct nuclear generating stations. A little less enthusiasm is called for; even a degree of reluctance to spread nuclear "benefits" more widely.[189] Commercial benefit to the exporting country should not be allowed to sway judgements on supplying nuclear reactors and fuel cycle facilities to such countries as Iraq, Pakistan, Syria, North Korea and Iran.

---

[189] Euratom, which began safeguarding inspections in 1959 and hence pre-dated the operation of the NPT, has a good record as a pioneer in this area of activity.

5. Nuclear Proliferation - the Hidden Costs

The profits to be made from selling nuclear plants to "doubtful" regimes are often large on an absolute scale, but trivial compared to the potential cost of a spread of nuclear weaponry. Paradoxically, in extreme circumstances the gift of a nuclear reactor, coupled to strict controls on the fuel cycle, can be used as a bribe in return for a country refraining from developing nuclear weapons. The US have adopted this course with North Korea and it has been suggested that, based on France's good relations with India, a deal might be struck whereby pressurised water reactors (PWRs) and technology could be exchanged for India abandoning nuclear weapon development. To the considerable expenditure of the US in North Korea must be added the large cost of maintaining a sizeable American military force in South Korea for very many years. In fact, the level of America's defence budget is predicated on the possible need to conduct two armed conflicts with recalcitrant states simultaneously, and a proportion of the cost of this hugely expensive policy arises from fear of the spread of nuclear weapons (and of course in about equal measure its need to preserve its supplies of oil from foreign sources).

America's willingness to finance these "hidden" costs of having to maintain very large conventional forces is in marked contrast to its Congress's recent parsimony when allocating resources to assist Russia's programme of disarmament. In the opinion of many observers, one of the greatest threats to world peace is the possibility of illicit diversion of fissile material from Russia's crumbling nuclear arsenals to terrorists or unstable and militaristic national states. If in recent years the US have become increasingly parsimonious, then Western Europe and Japan can only be described as recklessly irresponsible in relation to the trivial level of its assistance in helping Russia transform from a dictatorship with a huge nuclear arsenal to a democracy where nuclear weaponry plays a less dominant role in its national defence.

6. Reprocessing and Waste Management in Britain

Throughout the 1970s, it gradually became more apparent that the fast reactor could not become economically viable for many decades; experts began to talk about time-scales for their deployment starting half-way through the twenty-first century. In spite of President Carter's initiatives in banning US reprocessing, and the lack of a perceived use for the separated plutonium, British Nuclear Fuels Limited (BNFL) actively sought clients for its projected Thermal Oxide Reprocessing Plant (THORP) at Sellafield. Customers, principally Japan and Germany, were not hard to find because political pressures in these two countries were forcing them to do something with their spent fuel. Sending the spent fuel to Britain or France brought time, and to achieve this breathing space Germany and Japan were not only prepared to pay very high reprocessing charges but also, in effect, to pay an appreciable proportion of the capital costs of the plant.

An implied condition, as far as BNFL gaining foreign reprocessing orders were concerned, was that the single nationalised electrical utility serving all of England and

Wales, the Central Electricity Generating Board (CEGB), would also have to have reprocessed its Magnox and Advanced Gas Cooled Reactor spent fuel.[190] In spite of disquiet amongst some sections of the CEGB, and the fact that reprocessing charges would greatly exceed the cost of dry storage, BNFL managed to get government backing for its policy and the CEGB was "persuaded" to conform.

A consequence of this largely unnecessary reprocessing is that BNFL has now accumulated at Sellafield almost 50 tonnes of separated plutonium of British origin, much of which is derived from spent Magnox fuel. To this figure must be added some 35 tonnes of plutonium from foreign sources in spent fuel and already separated. The agreement with overseas customers is that the plutonium and highly active waste extracted from the imported spent fuel must be returned to the country of origin. Such transport of radioactive materials, and particularly plutonium, is of course highly undesirable.

In 1998, the distinguished nuclear analyst Professor Frank von Hippel suggested that instead of fulfilling its contracts to reprocess the spent fuel for its foreign customers, BNFL should simply store the fuel, i.e. *not* reprocess it, but instead, in due course, send to these customers the agreed amount of plutonium, but taken from its existing plutonium store.[191] It should do the same with the high level waste and thereby fulfil its contractual obligations to the satisfaction of its customers and at the same time keep its indigenous critics contented. This is certainly an ingenious idea but any scheme which might accelerate movement of plutonium from its very secure store at Sellafield to foreign parts would have to be viewed with a great deal of caution.

UK's store of plutonium has recently been studied by a Working Group of the Royal Society which has concluded that, at a time when the US and the Former Soviet Union (FSU) are taking strenuous efforts to convert their military plutonium to the "spent fuel standard", it is inappropriate for BNFL to add to its already large stockpile of separated "civil" plutonium. The Royal Society has recommended that the British government reviews the strategy and options for stabilising and then reducing the stockpile. The Group lists the options in the medium term as:

(a) adopting the once-through cycle for standard fuel (i.e. stop reprocessing);
(b) recycling Mixed Oxide (MOX) fuel in British and foreign reactors; and
(c) constructing a new British light water reactor specifically designed to burn MOX.

BNFL are keen to develop a MOX cycle, but there are difficulties. Firstly, because of the complexity of the fuel elements and the excessive handling necessary during manufacture, the Magnox and Advanced Gas cooled Reactors (AGRs) are not suitable for burning MOX and the UK's single Light Water Reactor (LWR), Sizewell B,

[190] To support this policy, the fiction was promulgated that the magnesium-clad metallic fuel from the Magnox stations could not be dry-stored.
[191] *Nature*, 394, 1998, p. 415.

can only operate with up to a third of its core loaded with MOX.[192] With only this capacity available it would take more than twice the design lifetime of Sizewell B to eliminate the plutonium stockpile. On the other hand, if Sizewell B were converted to take 100 % MOX loading, and two other LWRs of similar size and MOX burning capacity constructed, then the stockpile could be removed in about 20 years, a reasonable timescale.

Of course, the sale of MOX fuel abroad is another option but again there are problems. Un-irradiated MOX would be attractive to terrorists because plutonium could be extracted from it relatively easily by chemical means. Hence, a high level of security would be necessary both during transport of the un-irradiated fuel and during its storage at the site. Burning MOX is not popular with operators because the thought of burning plutonium in thermal reactors might well generate an adverse reaction from the public living close to the stations. The utilities in fact may well expect to be paid handsomely for burning MOX (this is certainly the case in the US). Even without such a charge by the utilities, MOX fuel is more expensive than conventionally produced fuel even when a zero value is placed on the separated plutonium. No doubt an expansion in the manufacture of MOX would lead to a reduction in cost, but it seems likely that, for the foreseeable future, the activity will have to be subsidised as a politically acceptable way of transferring separated plutonium, which already exists due to previous military or civil activity, to the "spent fuel standard". Such a subsidy would appear to be reasonable.

What seems entirely unreasonable is the reprocessing of spent fuel, in order subsequently to manufacture MOX using the separated plutonium. It has been suggested that such reprocessing is "spherically daft" - it is ridiculous whichever way you look at it. It should be emphasised that, during the irradiation of the MOX, as much plutonium is created (in the uranium oxide) as is destroyed by the fission process in the plutonium oxide. The starting and final product is essentially the same - plutonium protected from theft by being incorporated into a spent fuel standard. The whole tortuous procedure seems to represent a completely unnecessary complication which has arisen from a desperate attempt by the reprocessing industry, particularly in Britain and France, to stay in existence when the need for its services and products have disappeared. Of course, at the end of its cycle, the spent MOX fuel itself could be reprocessed and the derived plutonium recycled again, but this does not look to be an attractive procedure - it certainly leads to an undesirable build-up of other actinides.

There is another route for disposing of the civil (and military) plutonium and this consists of mixing it with high level waste and, with or without vitrification, depositing the mixture in a deep and inaccessible geological site. Although this disposal route has a great deal to recommend it, it is not being actively studied in the UK.

---

[192] The characteristics of plutonium fission reduces the efficiency of control rods designed for conventional uranium fuel and this restricts the amount of MOX fuel which can be loaded.

7. History of Nuclear Waste Policy in the UK

The study of nuclear waste disposal has a long and chequered history in Britain. Scientists at Harwell began research on vitrification of High Level Liquid Waste (HLLW) as long ago as 1958 and, by the early 1960s, a pilot scale process called FINGAL had been developed and "real" waste had been immobilised. However, in spite of having achieved a world lead in vitrification work, the programme was abandoned in 1966. In the early 1970s, the research was restarted and Harwell developed a scaled-up version of FINGAL, known as HARVEST. This was though a "batch" process and, when BNFL wanted to install a vitrification plant for their commercial reprocessing operations, they purchased the French continuous vitrification process known as AVM (*Atelier de Vitrification de Marcoule*). Both HARVEST and AVM used borosilicate glass because of its ability to dissolve large amounts of fission products and actinides and its relatively low softening temperature (though this entailed rather poor leaching characteristics).

In 1976, the public's vague concern about the nuclear waste issue was crystallised by the recommendations of a Royal Commission under the Chairmanship of Lord Flowers. The Commission's findings were published in a report called "Nuclear Power and the Environment", one of whose main conclusions was:

> "There should be no commitment to a large programme of nuclear fission until it has been demonstrated beyond reasonable doubt that a method exists to ensure the safe containment of long-lived, highly radioactive waste for the indefinite future."

As far as finding a final resting place for high level waste is concerned, the industry, taking up Flowers' recommendation, planned a series of test drillings at various sites throughout the UK, but these plans generated such a storm of local objections to the planning applications that the government of the day ordered it to be abandoned. In announcing the change in policy to Parliament in December 1981, the Minister for Local Government and Environmental Services quoted from the Second Report of the Radioactive Waste Management Advisory Committee (RAWMAC), which advocated that serious consideration be given to storing high-level waste at the surface in solid form for 50 years or possibly much longer.

Another outcome of the Flowers' Report was the creation by the nuclear industry and government of a separate organisation whose sole objective was to design and locate a site for the disposal of Intermediate Level Nuclear Waste together with a small quantity of Low Level Waste which had alpha-emitter levels too high for the Drigg near-surface disposal site. This organisation was named "Nirex" (Nuclear Industry Radioactive Waste Executive) and set about looking for suitable sites and attempting to carry out test drillings. These activities generated a huge backlash from the general public living in areas which were potential sites, and all planning applications were bitterly opposed.

In the period 1983-87, a number of locations for shallow disposition were investigated for short-lived Low Level Waste (LLW) and Intermediate Level Waste

(ILW), and by the end of this period four sites were under active consideration. Unfortunately for Nirex, each of these locations was inside a Conservative-held marginal constituency and four weeks before the 1987 General Election the government panicked and ordered that all four sites be withdrawn from consideration. This was an entirely politically-motivated act (it has notoriously become known as "The Four Site Saga") and from it stem many of the present difficulties.

In 1988, Nirex embarked on a national consultation with about 50 000 questionnaires circulated. Thousands of replies were received and the importance of retrievability was emphasised. The next stage was a programme of site-selection, following, as Nirex supposed, government policy. Starting with 500 sites, these were whittled down, by 1989, to 12 deep sites. Since each of these met the government guidelines on operational and post closure safety, selection was based on a wide range of safety and non-safety considerations: ease of transport, site accessibility, land ownership, national parks regulations, special rules covering areas of outstanding national beauty, environmental impact, public acceptability, etc. Attention became focused on two sites, Dounreay and Sellafield, and in 1991 Sellafield was chosen, i.e. a site close to the Calder Hall reactors, the Windscale piles and THORP, BNFL's Thermal Oxide Reprocessing Plant.

Of course, everyone recognised that it would be little short of a miracle if on the entire mainland of Great Britain the best geology for a nuclear depository happened to be at Sellafield, where much of the waste was produced and where 60 % of it was already located. This was not the point at issue. The decision was taken that, though not ideal, Sellafield met the government criteria, as far as was known at that time. There were also of course other advantages - transport of waste would be minimised, the area was used to being the home of a huge nuclear/industrial complex, and public opinion, though divided, was certainly more favourable to the construction than possibly anywhere else in the country (other than Dounreay where transport costs would be prohibitive). At the time the initial decision to focus on Sellafield was made, Nirex was under the clear impression that it had the support of the responsible Minister (Nicholas Ridley) and that of RAWMAC.

In June 1994, Nirex applied for permission to build an underground laboratory to investigate further the Sellafield site; this was to be known as the Rock Characterisation Facility (RCF). In December Cumbria County Council rejected the planning application, even though the laboratory was just an exploratory stage in the process. Nirex appealed and a public enquiry was initiated which started in September 1995. In late 1996, the Inspector reported. He supported the rejection and quoted such matters as the visual impact of the site and the parking of cars and possible damage to badger sets. In other words, he was applying the sort of planning criteria which would have been appropriate for a proposal to build, say, a small housing estate. The inspector also expressed concern at the choice of a coastal site because of fears that active species may migrate over millennia and be released into the sea. On such matters, which would in any case have been addressed by research in the proposed rock laboratory, it might be thought that the inspector was stepping outside his area of expertise.

On March 17th 1997, the Environmental Minister, John Gummer, announced that he supported the rejection of Nirex's application. In his pronouncements, which were very critical of the industry, the Minister appeared to have overlooked the fact that his department shared the responsibility for the evolution of the proposal. What was even more disturbing was that, the same day the Ministerial decision was promulgated, the Prime Minister announced the date of the next general election, just six weeks ahead. It might have been thought more prudent if the Minister had left this extremely important decision to be dealt with in less frenetic circumstances, and with longer deliberation, by his successor.

If the Nirex application was dealt with in both an inappropriate and some might argue, cavalier fashion, then it is of interest to examine how much all this has cost the industry and the country and to examine whether the issue raises matters of national and international importance. As far as cost is concerned, in getting to the point of making the application, which was the culmination of a long and painful journey, Nirex and the industry had spent something like £400 million, a not inconsiderable sum; the Planning Enquiry alone cost Nirex £20 million. Such enforced yet wasteful expenditures in Britain, and in other countries, has contributed to an ending of the development of nuclear electricity in the Western democracies.

Chapter 13 The Disposal of Separated Plutonium Stocks

by
Pierre Goldschmidt*

Introduction[193]

The issue of the safe disposal of plutonium has generated strong feelings and diametrically opposed solutions. Public concern about this most emotive subject tends to polarise the debate. In the attempt to explore rational solutions a number of weighty studies have been given wide publicity. We will refer to them extensively.

As underlined in the Foreword to the American Nuclear Society (ANS) Special Report on Protection and Management of Plutonium[194]: "Few events in history have borne more potential impact for the human condition than the discovery of plutonium. Though exceedingly rare in nature, this element has now been created in sufficient quantities to fuel not only powerful weapons but also nuclear reactors capable of providing sufficient energy to sustain civilisation for a millennium.

The end of the cold war marks a time for great rejoicing. It also marks the time for fundamental decision-making on how we assure that plutonium released from nuclear weapons programs is never again returned to this use. Further, how should civilian plutonium arising from nuclear electric power plants be treated?

Given its intrinsic power, it is not surprising that plutonium has attracted widespread interest throughout the world. However, it has also generated a great deal of misunderstanding. What is disturbing is that this misunderstanding is often found in the halls of government – the very places where decisions are made that can affect the quality of human life for centuries to come."

And in his preface to the same report, Nobel Laureate Glenn T. Seaborg writes: "In today's commercial nuclear power reactors that are already furnishing some 20 % of all the world's electrical energy, plutonium that is formed in the nuclear fuel assemblies supplies almost half of the energy produced. We know, too, that plutonium is the key to the long-term contribution of nuclear power to meeting the world's growing needs for energy. By allowing us to burn virtually all of the available uranium rather than just 1 %

---

* Pierre Goldschmidt is General Manager of SYNATOM, responsible for the nuclear fuel cycle management of Belgium's nuclear power plants. The views expressed in this chapter are his own.

193 The question of plutonium disposal has been the subject in recent years of a considerable number of extended studies by very competent and authoritative institutions and groups of experts. The present chapter is designed to make a coherent synthesis of the most important aspects of these studies and to put them into perspective.

194 American Nuclear Society (ANS) Special Panel Report, *Protection and Management of Plutonium*, Foreword byAlan E. Waltar, August 1995.

as we do at present, the use of plutonium makes nuclear energy by far the largest energy resource available; indeed, one that is virtually inexhaustible.

When I think back to the remarkable rate of discovery and the rapid accumulation of knowledge in those early days (of peaceful nuclear uses), I am troubled by how slowly much of this knowledge finds its way into the public consciousness and how much misinformation and misunderstanding arises and persists."[195]

This chapter is designed to present a series of facts and information on the need and the means to safely manage plutonium arising both from the dismantlement of nuclear weapons and from civil nuclear power activities, in the short and the longer term.

It is intended to give policy makers, the public and the media a better understanding of this important but controversial question and to recommend appropriate means for protecting and managing plutonium from all sources and locations.

1. The Issue

The issue of plutonium management is succinctly stated in the ANS Special Panel Report: "Plutonium presents two faces: one as an environmentally hazardous, nuclear explosive material, and the other as a unique energy resource. These contrasting features have given rise to conflicting views on how to deal with plutonium, ranging from its disposal to its use or preservation for use as a nuclear fuel. Because of this dual nature, the plutonium issue is often characterised in terms of whether plutonium should be viewed as a waste material to be disposed of, or as an energy resource to be used or conserved. (....) This is an oversimplified characterisation, since it overlooks the fact that disposal of plutonium as waste does not necessarily ensure its unavailability for weapons use."[196]

Interest in plutonium management has been greatly intensified in the recent past by the unprecedented agreements of the US and Russia to dismantle large numbers of their nuclear weapons.

The resultant release of important quantities of plutonium from weapons has given rise to major governmental programs in the US and Russia and to numerous studies in these two countries as well as in Western Europe, which can bring its considerable industrial experience in dealing with the civilian use of plutonium for electricity production to help solve the military plutonium problem. The question of what to do with plutonium is indeed a broad and long standing issue not confined to the dismantling of nuclear weapons.

However, before analysing the global situation, it is of great importance to distinguish the different types of plutonium and the forms in which they are presently stored.

Plutonium exists in essentially three forms: Weapons-Grade Plutonium; Separated Reactor-Grade Plutonium; and Reactor-Grade Plutonium in Spent Nuclear Fuel.

---

[195]ANS Special Panel Report, *Protection and Management of Plutonium*, Preface by Glenn T. Seaborg, August 1995.

[196] ANS Special Panel Report, *op. cit.*, p. 20.

1.1 Weapons-Grade Plutonium (WPu)

This plutonium has a high content - typically 93 percent or more - of Pu-239, the most suitable isotope for nuclear explosive use. It is usually produced in facilities specifically designed for this purpose but also, as a by-product, in some nuclear reactors which generate electric power.

To obtain this weapons-grade plutonium with a high Pu-239 content, uranium fuel elements have to be only slightly irradiated, i.e. during short periods of time - typically no more than a few weeks - in contrast to what is done in civil commercial nuclear power plants, where they are irradiated for 4 years or more.

To be usable in nuclear weapons, the weapons-grade plutonium must be extracted from slightly irradiated spent fuel elements by a chemical process termed "*reprocessing*". After extraction in the form of an oxide, the plutonium must be purified to a high degree and converted into plutonium metal, the form used in weapons.

Less than 8 kg of weapons-grade plutonium in metallic form is required for a nuclear weapon. Weapons-grade plutonium in metallic form must therefore be afforded the highest level of protection against theft or diversion.

1.2 Separated Reactor-Grade Plutonium

Reactor-grade plutonium typically contains less than 60 % of the isotope Pu-239 and more than 24 % of the isotope Pu-240, as well as higher isotopes that significantly complicate any contemplated use in a nuclear weapon and degrade its performance. Moreover, separated reactor-grade plutonium recovered from civil spent nuclear fuel is in oxide form. It is theoretically possible to make an explosive device with separated reactor-grade plutonium oxide, but in practice such a device would be too unreliable and unpredictable to be used as a nuclear weapon. Nevertheless, because of this theoretical possible use, separated reactor-grade plutonium oxide is often referred to as "*weapons usable*" material, a simplifying and therefore somewhat misleading characterisation.

1.3 Reactor-Grade Plutonium in Spent Nuclear Fuel

"Most of the plutonium produced to date in civil nuclear power reactors is still contained in the spent fuel assemblies in which it was produced. While in this form, plutonium cannot be used in nuclear weapons. It is also very dilute (approximately 1 percent of the spent fuel weight) and access to it can be obtained only after it has been separated from the associated highly radioactive fission products."[197]

Spent nuclear fuel is therefore considered by the US National Academy of Sciences to be the standard or paradigm of an effective proliferation-resistant form in

[197] *Ibid.*

which to confine plutonium. This level of proliferation-resistance has become known as the "spent fuel standard" and has been adopted by the United States Department of Energy (US DOE) as an objective for the form into which one should transform surplus ex-military plutonium before its disposal in an appropriate geologic formation.

The options available for managing weapons-grade and separated reactor-grade plutonium will be reviewed in the remainder of this chapter. The options available for managing spent nuclear fuel are reviewed and analysed in a separate chapter in this book.

2. Management of Weapons-Grade Plutonium

With the end of the Cold War, the conclusion of the Strategic Arms Reduction Treaties (START) and other agreements, the US and Russia started to significantly reduce their arsenals of nuclear weapons. As a result each side has accumulated large stockpiles of "surplus" weapons-grade plutonium (as well as of highly enriched uranium). Existing stocks of weapons-grade military plutonium have been estimated as indicated in table 1.[198]

Nuclear Weapons States	FSU[(a)]	US	France	China	UK	Total	International Safeguards
Inside weapons	35	31	2	2	1	71	No
Outside weapons	96	54	3	2	2	157	No
Total	131	85 [(199)]	5	4	3 [(200)]	228	No

In non-nuclear weapons states the total quantity of weapons-grade military plutonium is estimated to be 0.8 ton; Israel: 0.44 ton; India: 0.30 ton; N-Korea: 0.03 ton; Pakistan: ~ 0 ton.
[(a)]FSU: Former Soviet Union.

Tab. 1. Stocks of weapons-grade military plutonium (in tons, as of December 31, 1994). *Source: Plutonium and Highly Enriched Uranium 1996, D. Albright, F. Berkhout and W. Walker, op. cit.*

---

198 *Plutonium and Highly Enriched Uranium 1996*,World Inventories, Capabilities and Policies, D. Albright, F. Berkhout and W. Walker, Oxford University Press, 1997.

199 The total US Plutonium Inventory in September 1994 was 99.5 ton consisting of the 85.0 ton of weapons-grade plutonium mentioned in table 1, plus 13.2 ton of fuel-grade plutonium and 1.3 ton of reactor-grade plutonium; see *Plutonium: The First 50 Years*, US DOE, February 1996. Of the total of 85 ton weapons-grade plutonium, 38.2 ton have been declared excess to defence needs (as well as an additional quantity of 14.3 ton of military non-weapons-grade plutonium); cf. INFCIRC/549/Add. 6, 31 March 1998, and a few ton seem to have been placed under IAEA surveillance.

200 The total UK military inventory is about 11.8 ton plutonium (December 31, 1995) consisting of the 3 ton of weapons-grade plutonium mentioned in table 1, plus 8.7 ton of fuel-grade or reactor-grade plutonium. In its Strategic Defence Review published in the summer of 1998, the UK Government declared that 7.6 ton plutonium were held outside international safeguards. Much of this stock is no longer required for defence purposes and 4.4 ton plutonium (including 0.3 ton of weapons-grade plutonium) will be placed under EURATOM Safeguards; cf. Nuclear Fuel, December 14, 1998.

The US National Academy of Sciences considers that, "*the existence of this surplus material constitutes a clear and present danger to national and international security*". At first sight, it seems astonishing that material recovered from dismantled warheads in the US would constitute a greater danger to national and international security than the warheads themselves. In fact, the principal US concern is that Russian weapons-grade plutonium, if not securely disposed of, could be diverted to terrorist groups or to states aspiring to build nuclear weapons. This concern is based on the perception of a severe degradation of Russia's economy giving rise to an increase in unemployment of highly qualified scientists who might be tempted to work for groups or countries wishing to develop their own nuclear weapons.

Russia currently stores plutonium at a number of sites, many with inadequate security systems. There is concern that if security at these facilities is not improved, fissile material may be diverted to unauthorised states or terrorist organisations. As stated in a report to the United States Congress: "To decrease the likelihood of theft, the United States is paying half the cost of constructing a safe and secure storage facility for excess plutonium and highly enriched uranium at the Mayak site near Chelyabinsk. Completion of the facility's first phase is expected in early 1999, with the final phase a year later. Only excess material from dismantled nuclear weapons is to be stored there. Other fissile materials will continue to be housed at a variety of military and civilian facilities."[201]

It is generally believed that Russia will not reduce its stockpile of surplus weapons-grade plutonium unless the US engages Moscow in a co-operative and simultaneous disposal program in both countries. On 2 September 1998, President Clinton and President Yeltsin signed in Moscow a "Joint Statement of Principles for Management and Disposition of Plutonium Designated as No Longer Required for Defense Purposes" under which the US and Russia have agreed in principle to convert approximately 50 tons of plutonium, withdrawn in stages from nuclear military programs, into forms unusable for nuclear weapons.

2.1 Options available

Russian authorities have repeatedly indicated that for them plutonium is a valuable energy resource and not a waste product and should be used in the future for electricity production as Mixed Oxide (MOX) fuel either in their so-called VVER-1000 Light Water Reactors (LWRs) or preferably in their existing and future Fast Breeder Reactors (FBRs).

2.1.1 The US National Academy of Sciences Analysis

In 1995, the US National Academy of Sciences (NAS) concluded an extensive study on "Management and Disposition of Excess Weapons Plutonium" with the finding

[201] Congressional Research Service (CRS) Report for Congress, *Nuclear Weapons: Disposal Options for Surplus Weapons-Usable Plutonium*, May 1997, p. 8.

that "two options have emerged as the most promising ones for the timely disposal of excess weapons-grade plutonium to the spent fuel standard:

- the current-reactor/spent-fuel option would use light water reactors or Canadian deuterium-uranium (CANDU) reactors of currently operating types or evolutionary adaptations of them, employing MOX fuel in a once-through mode, to embed the weapons-grade plutonium in spent fuel similar to the larger quantity of such fuel that will exist in any case from ordinary nuclear electricity generation;

- the vitrification-with-wastes option would immobilise the weapons-grade plutonium together with intensely radioactive fission products in heavy glass logs of the type planned for use in the immobilisation of defence high-level wastes."[202]

Although the timing of all disposal options is subject to influences that are not entirely predictable, the NAS panel judges that:

"with a prompt decision to proceed in this direction - and given high national priority assigned to the task - fabrication of WPu-MOX fuel could begin in the United States as soon as the year 2001 (....). Most if not all of the 109 commercial LWRs operating in the United States in 1994 would be capable, without significant modification, of operating with at least one-third WPu-MOX fuel in their cores."[203]

"The earliest possibility for implementing the vitrification-with-wastes option in the United States would seem to be at the US Department of Energy's Savannah River plant, where vitrification operations to immobilise the defence high-level wastes (HLW) at that site are scheduled to begin at its "Defense Waste Processing Facility" (DWPF) in 1996. Although some technical issues require resolution before a decision to proceed can be confidently made, the NAS panel believes that the necessary preparations for adding weapons-grade plutonium to this process could be completed by 2005."[204]

The NAS panel concludes that the two approaches are comparable in security overall, that either would be adequate, and that no other known option is superior.

The panel also believes that the two options can meet all applicable environmental, safety and health criteria.

From an economic point of view, the panel considers that "*the monetary costs of alternative approaches to the disposition of weapons-grade plutonium are of secondary importance compared to the security aspects*" and that "*the lowest central estimate, at about $0.5 billion, is for the option using currently operating US LWRs that need no modification to use MOX safely*".

"Even the higher figure is probably less than what this weapon material once cost to produce, as well as much less than would be spent in the attempt to recover such material if it went astray. It is incomparably less than would be spent to try to deter or otherwise prevent its use in the form of a bomb in the hands of a potential adversary.

---

[202] National Academy of Sciences, *Management and Disposition of Excess Weapons Plutonium*, p. 3.

[203] *Ibid.*, p. 5.

[204] *Ibid.*, p. 8.

Thus, funding should not be allowed to become a barrier to carrying out plutonium disposition."[205]

And finally the NAS concludes that: "*The areas of uncertainty in the current-reactor/spent fuel option are primarily in licensing and public acceptance. Those in the vitrification-with-wastes option relate mainly to criticality and to the technical issue of mixing plutonium and high-level wastes.*"

2.1.2 The US Department of Energy's two-track decision

Confronted with these alternatives the US DOE decided to follow a two-track plutonium disposal plan. It has determined that between 8 and 17 tons of the US surplus plutonium is unsuitable for use as MOX fuel without extensive purification and will therefore be immobilised, and that about 35 tons of the Pu declared surplus could be used as MOX fuel in existing LWRs[206] and, after a once-through fuel cycle, the spent MOX fuel would be disposed of in a geological formation as envisaged for standard spent fuel.

This "two-track" Pu disposal plan is strongly criticised by those opposing the MOX-fuel alternative and favouring the immobilisation-with-wastes of all surplus weapons-grade plutonium.[207, 208]

The opponents claim that the MOX option will be seen as an encouragement for other national states to continue reprocessing spent fuel and that it would cost more since a significant part of the weapons-grade plutonium surplus is unsuitable for MOX fuel and has to be immobilised anyway.

On the other hand, those in favour of the MOX option argue the following:

- The immobilisation option does not degrade the isotopic composition of weapons-grade plutonium and is viewed by some (including Russia) as permitting easier re-incorporation into nuclear weapons as currently designed.

- "Plutonium is a valuable energy resource, not a waste material to be buried".[209]

- "The MOX option burns part of the weapons-grade plutonium and converts the rest to a safer form. Since all the plutonium would be irradiated on a once-through basis without reprocessing of the resultant spent fuel it does not constitute any form of US

---

[205] *Ibid.*, p. 12.

[206] As a preliminary step in procuring potential MOX fuel fabrication and reactor irradiation services from the private sector, DOE's Office of Fissile Materials Disposition released a "Program of Acquisition Strategy" (PAS) in July 1997. DOE requests that the irradiation of MOX fuel derived from weapons-grade plutonium begins no later than 2007 and ends in 2022. However, DOE would make at least 0.5 ton plutonium from weapons dismantlement available in 2001 to produce MOX fuel for lead test assemblies irradiation demonstration. See: *Commercial Nuclear Fuel from US and Russian Surplus Defense Inventories*, DOE/EIA, 0619, May 1998.

[207] "Bury the Stuff", Edwyn Lyman and Paul Leventhal, *The Bulletin of Atomic Scientists*, March/April 1997.

[208] "Burn it or Bury it ?", S. Dolley, Nuclear Control Institute, 28 March 1997.

[209] ANS Special Panel Report, *op. cit.*

endorsement of the broader use of reprocessing and plutonium in the civil fuel cycle."[210]

- "The anticipated costs of the Russian and US disposition programs, while measured in the low billions of dollars, are modest when one considers the important security benefits that will be gained by achieving the spent fuel standard. Also, the likely cost and lead times of the MOX fuel and immobilisation alternatives are roughly comparable, and the costs of pursuing two tracks in some combined manner are only incrementally more expensive than the cost of using just one approach."[211]

- "There is much reason to think that the Russians will not eliminate their plutonium stockpiles at all if the United States implements only immobilisation, leaving all US plutonium weapons-grade - the Russians might then merely store their stockpile of weapons indefinitely, which is what we should most wish to avoid."[212]

In order to accelerate the disposal of excess weapons-grade plutonium, Senator Pete Domenici (Republican, New-Mexico) has been a strong supporter of the idea of burning MOX-fuel, made from surplus US weapons-grade plutonium, in European reactors.

A similar proposal was made a few years ago by the US "Natural Resources Defense Council" (NRDC): "While NRDC favours the vitrification option, last November (1994), we suggested an alternative MOX option for consideration, whereby the US Government would offer to supply fresh MOX fuel to foreign reactor operators in exchange for an equivalent quantity of spent fuel, at a price equivalent to, or below, the sum of the current market prices for reprocessing and MOX fuel fabrication. This offer would be directed primarily at Japanese and European utilities that are likely to contract with COGEMA (*Compagnie Générale des Matières Nucléaires*, France) and BNFL (British Nuclear Fuel Ltd, UK) for a second round of spent fuel reprocessing. In effect, the US would compete directly with BNFL and Cogema for spent fuel management services, but would provide these services at an equivalent or reduced price without actually reprocessing any spent fuel. Under this option, DOE in theory at least could actually make money disposing of its excess plutonium."[213]

However, in a July 14, 1998 letter to President Clinton's National Security Adviser, Senator Domenici wrote that on a recent European tour he did not find "a receptive audience" that would consider introducing weapons-grade plutonium into the European Nuclear Economy. According to Senator Domenici, the Europeans said adding weapons MOX would "upset their goal of balance within their civilian nuclear cycle between plutonium recovered from spent fuel and plutonium expended as MOX fuel".[214]

---

[210] Center for Strategic and International Studies (CSIS), *Disposing of Weapons-Grade Plutonium*, March 1998, p. 5.
[211] CSIS, *ibid.*, p. 5.
[212] CRS Report for Congress, *op. cit.*, p. 18.
[213] *Dismantlement of Nuclear Weapons and Disposal of Fissile Material from Weapons*, Thomas B. Cochran, July 1995.
[214] Nuclear Fuel, Vol. 23, n° 15, July 7, 1998, p. 3.

As we shall see from the analysis of the European situation, this reaction is quite logical. What is less understandable is the apparent rejection by the USDOE of the possibility to have lead-test MOX fuel elements fabricated in Europe with US plutonium and loaded in US nuclear reactors, in order to accelerate the MOX licensing procedure in the United States and thereby the disposal of surplus US's weapons-grade plutonium.

2.2 Facing Reality

From what has been said and written over the last five years, it is clear that Russia (a) considers weapons-grade plutonium to be a resource to produce electricity and not a waste, (b) has no intention to recycle weapons-grade plutonium in existing Russian VVER-1000 as long as it is not economically justified. Since Russia has large highly enriched uranium (HEU) stockpiles that can be blended down to low enriched uranium (LEU) to fuel existing reactors and has an overcapacity in uranium enrichment services, using weapons-grade plutonium as MOX fuel is not competitive in the short term unless the MOX-plant to be constructed is completely financed by foreign investors.

In fact, Russia is not worried about its stockpile of weapons-grade plutonium, still produces weapons-grade plutonium in its so-called RBMK reactors, and appears to be in no hurry to diminish or eliminate its plutonium stockpile unless there is a clear economic advantage to do so as was the case with the sale of part of its surplus HEU stock to the United States.

The US, after several years of effort, is still primarily in a study-and-evaluation stage in determining the next steps. There are grounds for serious concern about whether the United States is devoting attention and resources sufficient to assure the success of the disposal process.

On September 2, 1998, Presidents Clinton and Yeltsin signed in Moscow a Joint Statement confirming the intention of each country to remove in stages about 50 tons of plutonium from nuclear weapons programs and to ensure that the material could never be used for military purposes. However, the Joint Statement recognises "that in order for this effort to be carried out, it will be necessary to agree upon appropriate financing arrangements."

Indeed, Russia is reportedly seeking the equivalent of some one to two billion dollars from the international community to finance its work on the joint project.

As stated by John P. Holdren[215] in 1996: "The costs to the United States of improving the protection of nuclear materials in the former Soviet Union should be seen as an investment in national and international security, just as the cost of producing the US stockpile of nuclear weapons and weapons materials was once viewed. The cost of failure to act - in higher defense budgets and lower security in the future - would be far higher than the cost of timely action now."

---

[215] "Reducing the Threat of Nuclear Theft in the Former Soviet Union", *Arms Control Today*, John P. Holdren, March 1996.

But a March 1998 report of the Center for Strategic and International Studies (CSIS) takes the position that "the United States cannot be expected to finance its own disposition program and the bulk of the Russian disposition program as well."[216]

One must question whether such a position undermines the credibility of the US concern about excess military plutonium stockpiles and its willingness to solve the problem rapidly, the more so when, at the same time, the US DOE is considering spending more than $10-billion to construct and operate an accelerator for the production of tritium[217], a radioactive gas used to boost nuclear weapon's explosive power[218] and $1.8 billion for emergency security improvements at most of the US embassies and consulates world-wide against terrorist actions.[219]

2.3 Priority actions

As explained in the "Joint US/Russian Plutonium Disposition Study"[220]: "Neither the US nor Russia conceives of storage as the ultimate solution to plutonium disposition. In both countries, however, a substantial period of plutonium storage is a necessary first step for all options, as none of the plutonium disposal options can be implemented immediately. Providing safe, secure, and inspectable (verifiable) storage for all excess fissile materials is an urgent task. Ensuring safe and secure transportation of fissile materials to and from the storage facilities is also critical.

The United States and Russia are co-operating in the design and construction of a modern, safe, and secure storage facility in Russia for fissile materials from dismantled nuclear weapons. This is because existing facilities do not have sufficient capacity and do not meet present safety and security standards.

Construction of the new facility is underway at the Mayak complex at Ozersk. (...) The facility's operation will require approximately 200 full-time personnel, and is designed to last 100 years. Current estimates, which are to be updated as more detailed designs are completed and analysed, are that the facility will open in 1998 and will cost over $300 million, to be split roughly evenly between the US and Russian sides."

From an objective analysis of the present situation concerning excess plutonium from dismantled warheads, it is clear that support should be given to the following recommendations made in the Center for Strategic and International Studies (CSIS) 1998 Report:

"Without detracting in any way from the major goal of actually achieving the spent fuel standard as soon as possible, policy-makers need to devote much greater attention to two important objectives along the path.

---

[216] CSIS, *op. cit.*, p. 7.

[217] Or between $ 1.35 and $ 1.9 billion in up-front costs to the Government if Tennessee Valley Authority reactors are used to produce tritium, cf. Nuclear Fuel, December 14, 1998.

[218] Nuclear Fuel, Vol. 23, n° 14, July 1998.

[219] *International Herald Tribune*, September 23, 1998.

[220] *Joint US/Russian Plutonium Disposition Study*, Summary Report, August 1996, p. 4.

• First, as a matter of high priority, the two countries should start converting their excess stocks that are now in classified shapes (such as plutonium pits) to unclassified forms. The existing materials in these classified shapes can readily be recycled into nuclear weapons and cannot be placed under traditional International Atomic Energy Agency (IAEA) safeguards without compromising sensitive weapons information.

• Second, in parallel with concrete on-the-ground efforts to reduce the vulnerability of fissile materials, we (the US) should pursue verified and early legal commitments against fissile-material reuse in nuclear weapons. These commitments may be achievable sooner than any excess plutonium stocks can finally be disposed of, but they should not be pursued to a degree that will slow down concrete actions that would reduce the vulnerability of these materials.

By giving near-term priority to these two specific topics, the United States and Russia will be demonstrating their respective commitments to advancing the principle of irreversibility and will be achieving demonstrable advances in reducing the nuclear threat.

It also will be important to assure that strict non-proliferation controls and conditions continue to apply throughout the entire plutonium-disposal process and that they remain fully credible for both the immobilisation and MOX fuel approaches. It will also be vital that the United States and Russia continue to collaborate closely with the IAEA to ensure that their excess stocks of weapons-grade plutonium are placed under effective verification measures as soon as practicable. It will be important, in this regard, to ensure that effective verification measures are endorsed early in the program and that they are integrated from the start in the design of any relevant new facilities."[221]

3. Management of Separated Reactor-grade Plutonium

Civilian reactor-grade plutonium is stored primarily in two forms:

- Pu oxide (PuO_2) separated after reprocessing spent fuel;
- PuO_2 contained in spent fuel.

There is an intermediate category which consists of the PuO_2 that is contained in fresh MOX-fuel elements (see § 4.4.3 below) awaiting their irradiation in nuclear power plants. This intermediate category does not give rise to any different security or environmental concerns.

[221] CSIS, *op. cit.*, p. 7-8.

3.1 General Background

In the 1960s, a number of European countries decided to have their nuclear spent fuel reprocessed in order to recover fissile materials (U and Pu) as fuel for future reactors. These decisions were made in the light of the then prevailing assumptions about global resources of uranium ore and its production cost, the prospects of building a series of fast breeder reactors using plutonium fuel, the necessity to transform fission products and other high level wastes into an appropriate form for long term storage and disposal, and the role of nuclear electricity production in national energy programs, as well as prevailing national regulations. In Germany, for instance, until 1994, the atomic law required electric utilities to reprocess all their spent fuel.

These decisions led to the construction of a number of reprocessing plants in the UK, France, Germany (a project abandoned in 1989) and, for a time, in Belgium, where the internationally owned Eurochemic reprocessing plant operated from 1966 to 1974. Most of the plutonium recovered in the 1960s and 1970s was used to fabricate the first cores of the European fast breeder reactors (SNR-Kalkar in Germany, Phenix and Superphenix in France, and PFR-Dounreay in the UK).

In this context, a number of very long term reprocessing contracts were signed, on a cost plus fee basis, by many European and Japanese utilities with the two large European reprocessors: BNFL and COGEMA. These contracts, executed in the 1970s, are still running and new contracts have been signed since then for the post 2000 period.

Many assumptions made 20 years ago are no longer valid today. The question of whether reprocessing of spent fuel is still justifiable is analysed in the preceding chapters dealing with spent fuel management.

The fact that the development of fast breeder reactors in Europe has been much slower than initially anticipated and more recently almost abandoned (at least for the time being), has led to an imbalance between the separation of plutonium under existing reprocessing contracts and its use in FBRs. Some utilities therefore decided to burn this excess plutonium as MOX-fuel in existing LWRs.

However, up to now, MOX-fuel fabrication capacity has not been sufficient to recycle all the plutonium separated under the existing reprocessing contracts.

Not the least of the reasons for this lack of MOX-fabrication capacity is the strong opposition of anti-nuclear and green forces which, among other things, succeeded in having the 1.2 billion DM Hanau MOX-fabrication plant abandoned in 1991. The result is that many of the utilities which signed reprocessing contracts with BNFL and COGEMA currently have some excess reactor-grade plutonium.

3.2 The UK situation

The UK is in a special situation because its first 26 nuclear reactors (of which 20 are still operating) were of the Magnox type (magnesium-clad gas-cooled), operated by government-owned entities. For technical reasons, spent Magnox fuel (in contrast to LWRs fuel) can not be stored as such for an extended period of time. It was therefore decided very early on to reprocess that fuel. A Magnox reprocessing plant has been operating since 1964 and is expected to continue operation through to the end of the Magnox program.

Thereafter, it was decided that the spent fuel from the 14 succeeding reactors, which are of the AGR (Advanced Gas-cooled Reactor) type, presently operated and owned by British Energy, would also be reprocessed. There are currently no plans to reprocess the spent fuel from Sizewell B, the only LWR in the UK, none of the present generation of UK nuclear reactors is licensed to use MOX fuel and the only good candidate for doing so seems to be Sizewell B. Historically, the UK policy was based, as in other European countries, on the assumption of commercial use of plutonium in fast breeder reactors. This has not materialised as forecast. In 1990 the UK government decided to phase out its support of the European Fast Reactor Project, and in 1994, effectively withdrew its support of the UK prototype fast reactor at Dounreay.

This situation has inevitably resulted in the build up of significant separated plutonium stocks. Current stocks of separated civilian UK plutonium amount to 53.5 ton (most of it originating from the Magnox reactors) and is forecast to rise to over 100 ton by 2010.

3.3 The Belgian situation

The Belgian situation is worth examining here because it lies at the other end of the spectrum from the UK situation.

As indicated earlier, European and Japanese utilities with significant reprocessing commitments have, to varying degrees, some backlog of separated plutonium that they will only progressively be able to recycle as MOX fuel in their nuclear power plants.

Presently, Belgian utilities seem to be the only ones that have signed significant reprocessing contracts in the 1970s and are nevertheless not faced with any excess plutonium stockpile. Plutonium arising from their first three reprocessing contracts has been sold or leased with the agreement of the Euratom Supply Agency to companies involved in the European fast breeder reactor programs and MOX fuel fabrication.[222] The destination of the plutonium arising from the Belgian utilities' fourth and largest reprocessing contract was debated for an entire year in the Belgian Parliament, resulting in the Belgian Government's 1993 determination:

- that existing reprocessing contracts could be fulfilled and the resulting separated plutonium recycled as MOX fuel in two of the country's large Pressurised light Water Reactors (PWRs); and
- that the reprocessing/recycling option and the once-through cycle option should, from there on, be considered on an equal footing.

---

[222] "La gestion du combustible usé en Belgique et l'utilisation du combustible MOX dans les centrales belges", Ministère des Affaires Etrangères, Ministère de l'Emploi et du Travail, Ministère de la Santé Publique et de l'Environnement, Octobre 1992.

This determination made it possible for Belgium's largest utility, ELECTRABEL, to effectively reload the first MOX fuel elements in its Tihange 2 and Doel 3 nuclear plants in 1995. By limiting the MOX fuel elements to 23 % of the total number of fuel elements in the reactor core, no modifications to these plants nor to their operating mode were needed. The plan will enable Belgium to recycle all its plutonium recovered from reprocessing (4.8 tons in total) by the end of 2003.

The MOX-fuel assemblies loaded in the Belgian reactors of Doel 3 and Tihange 2 are designed to reach the same burn-up as standard enriched uranium fuel assemblies.

3.4 Separated civilian plutonium stocks

A global picture of the situation at the end of 1993 can be found in the book "Plutonium and Highly Enriched Uranium 1996".[223] The situation at the end of 1997, based on information transmitted by Belgium, China, Germany, Japan, the Russian Federation, the UK and the US have been published by the International Atomic Energy Agency, under the acronym "INFCIRC/549". These data can be summarised as shown in table 2.

3.5 The Alternatives

European and Japanese entities owning civilian reactor-grade plutonium stocks have to decide on the best way to dispose of this material. The available alternatives are similar to those considered by the US for their excess military plutonium: first safe storage and then, either separated plutonium is used as a fuel to produce electricity, or it is treated as waste. In February 1998, the UK Royal Society published a very thorough analysis[224] of the technical options, currently available or under development, for managing or using the UK's separated plutonium stock. Although the UK situation is the most extreme in Europe, it is worth examining the Royal Society's findings which can be summarised as follows.

---

[223] *Plutonium and Highly Enriched Uranium 1996, op. cit.*

[224] The Royal Society, *Management of Separated Plutonium*, February 1998.

Country of Ownership		In Store *31-12-1993* (a)	IAEA – INFCIRC/549 In Store *31-12-1997 (1996)*		IAEA – INFCIRC/549 In Fresh MOX Fuel *31-12-1997 (1996)*		International Safeguards
NNWS	**Euratom**	13.5[1]	2.6[2][5]	(1.8)	7.0[2]	(5.8)	Eur+IAEA
	Switzerland	0.2	0.1	(0)	0.6	(0.1)	IAEA
	Japan	10.7	N.A.	(16.1)	N.A.	(4.0)	IAEA
	Sub-total	24.1	N.A.	(17.9)	N.A.	(9.9)	-
NWS	**China**	-	-	-	-	-	No
	France	15.6	20.2	(19.1)	18.5	(16.3)	Eur+IAEA
	Russia	26.5	27.2[3]	-	0.1[3]	0.1[3]	No
	UK	39.2	52.2	-	2.7	(2.7)	Eur+IAEA
	US	1.5	1.7[4]	1.7[4]	0[4]	0[4]	No
	Sub-total	82.8	101.3	-	21.3	(19.1)	-
	India	0.3	N.A.		N.A.		No
Grand total		107.2	≥138[6]	-	≥32.9[7]	(29)	-

NNWS: Non Nuclear Weapons States; NWS: Nuclear Weapons States; Eur: Euratom; N.A.: not applicable.
(a) *Source: Plutonium and Highly Enriched Uranium 1996, D. Albright, F. Berkhout and W. Walker, op. cit.*
(1) Euratom's NNWS considered are : Belgium, Germany, Italy, the Netherlands.
(2) Euratom's NNWS considered are : Belgium and Germany.
(3) As of 1-7-1996.
(4) Quantity of US civilian industry fuel-grade plutonium (in September 1994) that was acquired by the US Government (*Plutonium : The First 50 Years*, US DOE, February 1996).[225]
(5) In INFCIRC/549/Add. 2/1, Germany did not indicate the quantities of plutonium in store in other countries at the end of 1997.[226]
(6) Assuming 16.1 ton plutonium for Japan, as of 1996, and an additional 18 ton plutonium for Germany.
(7) Assuming 7 ton plutonium for Japan, as of 1996.

Tab. 2. Civilian plutonium separated after reprocessing (in tons Pu total).

3.5.1 Storage

"There appears to be no fundamental technical reason to change from the current practice of storing separated plutonium. Plutonium stores are designed to be safe for 50 years and it is possible that this could be extended. The method is well proven and economic, and facilities are already in place, so this is a pragmatic immediate solution for

[225] The figures of 40.4 ton plutonium in store and 4.6 ton plutonium in fresh MOX fuel given, as of 31 December 1996, in INFCIRC/549/Add. 6, as of 31 March 1998, do not seem to be correct and have not been reported here.
[226] It is estimated that 22 ton of fresh separated plutonium of German origin are stored at the reprocessing plants of La Hague and Sellafield, at the end of 1998. Therefore, at the end of 1997, an estimated quantity of 18 ton plutonium should be added in this table to the 2.6 ton plutonium figure.

plutonium that has already been separated. This policy does not prevent any other option being taken up in the future."[227]

However, this "stockpile can be viewed as a strategic and environmental risk, as well as an open-ended legacy for future generations." Therefore, "Safeguards and non-proliferation measures should remain under continuous review."

3.5.2 Immobilisation and Disposal

"It could be possible to dispose of separated plutonium, after suitable pre-treatment, by mixing it with highly radioactive waste and either converting the mixture into glass by the vitrification process or immobilising the plutonium in ceramic pellets. This process does not extract useful energy from the plutonium and does not destroy it, but renders it relatively inaccessible. The product could be disposed of in geological structures. Such technologies exist for immobilisation of HLW but different equipment designs and materials formulations will be required because of chemical differences between plutonium and high level waste, which may affect the stability of the glass or ceramic matrix.

In addition to the technical uncertainties, geological disposal calls for a new approach to safeguards. A major quantity of plutonium would be underground and require long-term safeguards maintenance. It is not clear how the financial and legal arrangements for such permanent supervision can be established. A disadvantage of disposal is the waste of a potential fuel resource."

"In view of the difficulties currently being experienced by plans for deep disposal of low and medium level radioactive waste, it is hard to imagine deep disposal of plutonium being a viable option in the medium term."[228]

3.5.3 The MOX option

"Plutonium oxide can be mixed with uranium oxide to form MOX fuel for use in the present generation of thermal reactors. Such use of civilian MOX fuel in thermal reactors would reduce the amount of separated plutonium in store and convert it to a form consistent with the spent fuel standard, but would not eliminate it as long as U is the main component of the reactor core."[229]

This option is well proven technically, with experience of loading MOX fuel into a thermal reactor dating back to 1963 (BR-3 reactor, Belgium). Large-scale use commenced in Germany in 1981 and the government-owned French utility started to recycle plutonium in thermal reactors in 1985.

"The economics of MOX recycling in thermal reactors are controversial. Reprocessing of spent fuel solely to produce plutonium for recycling as MOX is not economic. But given that reprocessing has been carried out and the plutonium is available, then the extra cost of fabricating MOX fuel (as compared with enriched

---

[227] The Royal Society, *op. cit.*, p. 7.
[228] *Ibid.*, p. 9.
[229] *Ibid.*, p. 8.

uranium fuel) might be justified in view of the savings in uranium and enrichment costs."[230]

3.5.4 Utilisation of UK MOX fuel in overseas reactors

"At present no UK reactors are licensed for burning MOX. Sizewell B, the LWR, could utilise MOX fuel." (....) "While there do not appear to be any insurmountable technical difficulties with loading MOX into AGRs, doing so would present a number of practical difficulties." (....) "Magnox reactors have metallic fuel and incorporation of plutonium into them would present formidable technical and licensing difficulties which would not be worthwhile pursuing for a reactor system of such limited remaining life."

"Even if these options were brought into use in the UK, a substantial plutonium stockpile would remain. UK plutonium could also be "leased" or sold to overseas utilities as MOX fuel. (....) The MOX would be under international safeguards and under Article 4 of the Non-Proliferation Treaty it would be subject to end-use restrictions, meaning that it could not be used for the development or production of nuclear weapons. The US is generally opposed to a "plutonium economy" and so may not be willing to take part in such transactions. However, France, Germany, Belgium and Switzerland all have reactors licensed for MOX, with spare capacity. Belgium in particular has reactors licensed for burning MOX, but has a moratorium on (new reprocessing contracts) in place and so spare capacity for burning MOX[231] would appear to exist."[232]

3.5.5 Alternative fuels and reactor designs

In the longer term, alternative fuel cycles and advanced reactor designs may make it possible to burn stockpiled plutonium more efficiently, or, in some configurations, to generate nuclear energy without producing any plutonium at all. Research and development capabilities should be maintained in OECD (Organisation for Economic Co-operation and Development) countries to allow evaluation of competing systems and support international collaboration on those likely to be of benefit to future generations.

4. Separated Plutonium Management: Comparing the situation in the US and in Europe

At this point it may be interesting to compare the situation created by the build-up of plutonium stocks in the US and in Europe.

---

[230] *Ibid.*, p. 8.

[231] Note: Loading fabricated MOX-fuel elements from 2003 on.

[232] The Royal Society, *op. cit.*, p. 9.

4.1 Plutonium Storage

• In the US, as in Russia, *weapons-grade* plutonium from dismantled nuclear warheads is initially in the form of "pits". For reasons of security, the plutonium *metal* pits must be transformed into an oxide form before either disposal option can proceed. The US does not currently have a pit disassembly and conversion facility capable of operating on an industrial scale.

The total investment and operating cost of converting 50 tons of weapons-grade plutonium to an oxide form in the US is estimated by the US DOE to amount to 1,330 million of 1996 dollars[233] (i.e. $26.6/gPu to be compared to $7/gPu mentioned by the US NAS as the cost of converting plutonium metal to oxide[234]). US DOE estimates that adverse variants in the separation process, such as gallium removal, could even add an additional $200 million.

• By contrast, in Europe, all *reactor-grade* plutonium separated after reprocessing spent fuel is already in the form of plutonium *oxide* of fabrication quality.

Storage costs per se are probably similar in the US and Europe for plutonium oxide. However, all civilian separated plutonium stored or used in European facilities are subject to both Euratom and IAEA safeguards. To date, excess weapons-grade plutonium stocks are still not subject to international safeguards in the US (nor in Russia).

4.2 MOX fabrication

In the US, no MOX fabrication facility exists. In Europe, on the contrary, there are four operating plants:

- One plant in Belgium: the BN-Dessel fabrication capacity (35 t HM/y)[235], in operation since 1984;
- Two plants in France:

-The "Centre de Fabrication de Cadarache" (CFCa), which has been processing fuel since 1962, mainly for fast breeder reactors. Since 1989, it has produced about 20 t HM/y of MOX fuel for light water reactors, reaching a capacity of 30 t HM/y in 1995;

-The MELOX plant in Marcoule, which started operation in 1995 and has a capacity of 100 t HM/y that could be expanded to 160 t HM/y or even more;

- One plant in the UK: the MOX Demonstration Facility (MDF) at Sellafield operational since 1993 (8 t HM/y).

Also, a new large fabrication plant (120 t HM/y), the Sellafield MOX Plant (SMP) has been constructed in the UK and is awaiting government authorisation to start up.

[233] CRS Report for Congress, *op. cit.*, p. 14.

[234] National Academy of Sciences, *op. cit.*, p. 302.

[235] The unit tHM means "ton of heavy metal", i.e. the weight of uranium and plutonium metal.

Together, these five plants are capable of transforming each year about 20 tons of plutonium into MOX-fuel. To put things into perspective, such a capacity is sufficient to convert all US surplus weapons-grade plutonium into MOX-fuel in less than 3 years.

4.3 MOX-fuel irradiation

- In the US, according to the National Academy of Sciences: "Most if not all of the 109 commercial LWRs operating in 1994 would be capable, without significant modification, of operating with at least one-third WPu-MOX fuel in their cores." Only 7 percent of the US LWR capacity would be needed in order to irradiate 50 ton weapons-grade plutonium fabricated into MOX over a period of 25 years. Moreover, just two 1,200 MWe-class PWRs (such as the three Palo Verde reactors operating in Arizona), using 100 percent MOX cores, could irradiate 50 ton weapons-grade plutonium in 25 years and thereby reach the spent fuel standard.

The task, therefore, seems quite modest. Even though some reactors will not require modification in order to load MOX-fuel, some experts[236] believe that the costs for MOX-related safety analyses and licensing would amount to $100 million. US DOE has also budgeted an additional $100 million for delays in reactor modification to accommodate MOX-fuel. Given potential public opposition to utilities handling plutonium in civilian reactors, US companies that have expressed interest in burning MOX fuel have done so on the assumption that the US government would provide substantial subsidies estimated by US DOE at $500 million.[237]

- In Western Europe, most of the 115 operating LWRs would be capable, without significant modifications, of operating with at least one third MOX-fuel. Among these, 30 are already burning MOX-fuel (2 in Belgium, 16 in France, 9 in Germany and 3 in Switzerland), 9 more have obtained the license to do so (4 in France, 5 in Germany) and 8 additional licences have been requested.

The accumulated experience with MOX-fuel in Europe represents more than 40 reactor-years.[238] No difficulty has been experienced in meeting the safety criteria nor the operational requirements. None of the fuel rods loaded in light water reactors ever failed due to a cause linked to the MOX design or fabrication.[239]

To date, European utilities have only burned self-produced civilian reactor grade plutonium, i.e. recovered after reprocessing their own spent fuel. Since, in Europe, recycling separated civilian plutonium as MOX-fuel is entirely unsubsidised by governments, the European utilities have clearly chosen the MOX-option as the most economical, as well as the only technical solution available today on an industrial scale, to achieve the spent fuel standard with separated plutonium.

---

236 "The 'Value' of Excess Weapon Plutonium", Richard Garwin, Amaldi Conference, October 1995.

237 CRS Report for Congress, *op. cit.*, p. 15.

238 "MOX fuel fabrication, in reactor performance and improvement", J. Van Vliet, P. Deramaix, J. L. Nigon, W. Fournier, ENC'98, Nice, 25-28 October 1998.

239 "MOX fuel fabrication, in reactor performance and improvement", *ibid.*

- In Japan, the Tokyo Electric Power Company and the Kansai Electric Power Company will commence MOX utilisation in 1999 in two plants and two more in the year 2000. It is expected that by the year 2010, the number of nuclear power plants loading MOX-fuel will be between 16 and 18.[240]

4.4 Economic evaluation

In light of these considerations, it is easily understandable that the economic consequences of recycling separated plutonium as MOX-fuel in LWRs are significantly different in the US and in Europe.

4.4.1 In Europe

Western Europe, after some 35 years of MOX-fuel utilisation, currently has large scale MOX-fabrication plants operating on an industrial basis, and has demonstrated the excellent performance of MOX-fuel in LWRs. All these operations are subject to the most stringent international non-proliferation controls (both Euratom and IAEA safeguards). Western Europe has also developed, over more than 30 years, appropriate techniques for HLW vitrification on an industrial basis. It is in fact currently exporting its know-how to the US for the clean-up of the Hanford military site.

However, these HLW vitrification processes have targeted the lowest possible plutonium content and are not readily applicable to Pu vitrification without time-consuming and costly R&D efforts.

Some experts have attempted to demonstrate that recycling separated plutonium as MOX-fuel in Europe is not economically viable. These calculations usually compare all the additional costs associated with the fabrication, licensing and use of a fresh MOX-fuel element, to the cost of a fresh standard uranium fuel element, and conclude that the value of plutonium is negative. This approach fails to take into account the fact that the already separated plutonium, if not recycled in MOX-fuel to produce electricity and reach the spent fuel standard, has to be stored, immobilised and disposed of as waste and could therefore have a significantly greater negative value.

In fact most European utilities capable of loading MOX-fuel in their LWRs have found, based on market-related commercial offers from suppliers, MOX-fuel to be the cheapest option to reach the spent fuel standard. Otherwise, in an increasingly deregulated electricity market, it is doubtful that they would make such a choice.

A special case is worth mentioning. Due to the size of the UK civilian separated plutonium stock and the limited possibilities of recycling it in the UK's nuclear power plants, the Royal Society has suggested that "UK plutonium could also be leased or sold to overseas utilities as MOX-fuel." Selling the UK's plutonium overseas does not seem politically acceptable today, even if it were economically viable. But the concept of "leasing" UK's plutonium to overseas European utilities is worth consideration. The UK

[240] IAEA, INFCIRC/549/Add. 1, 31 March 1998.

owner of the plutonium would have it fabricated as MOX-fuel and delivered to a European utility for loading in a LWR. The utility would pay for the energy produced from the MOX-fuel elements and, after an appropriate cooling time, would return the spent fuel to the owner of the original plutonium and receive in exchange, at an agreed time, a quantity of vitrified HLW corresponding to that contained in the spent fuel. This concept should be politically acceptable since the owner of the separated plutonium would take back a smaller quantity of plutonium in the form of spent fuel, and the utility would take back waste equivalent to that which it had created by the production of electricity. The question is then to find out whether the negative value assigned to excess separated plutonium by the UK owner is sufficient to make the deal attractive to both the plutonium owner and the utility (see Appendix A for a numerical example). Such a concept should be supported by those opposing reprocessing (because it would displace the possibility of recycling plutonium from new reprocessing contracts) and by those claiming to be worried about existing plutonium stockpiles.

Similarly, Frank von Hippel has recently suggested[241] that Britain could trade some of its stockpile of separated reactor-grade plutonium and associated radioactive waste for unreprocessed foreign spent fuel, implying thereby that this plutonium would be used as MOX fuel in Europe and Japan.

4.4.2 In the United States

In the US the situation is completely different, because no industrial facilities exist, either for MOX-fuel fabrication or for weapons-grade plutonium immobilisation. Economic comparisons can therefore only be made on investment and operating cost assumptions, and not yet on commercial offers. When one considers small or large MOX fabrication plants with different operating lives, the uncertainties on estimated US licensing costs and possible delays due to lack of public acceptance, and estimates of the level of subsidies likely to be requested by the US utilities to use MOX-fuel, it is not surprising that the economic calculations done by different sources can lead to widely varying results. To date, there has not been a convincing demonstration that treating weapons-grade plutonium as a waste in the US would be more advantageous than recycling it as MOX-fuel for the production of electricity in existing LWRs, and one is tempted to believe that the latter option would constitute a win-win-win solution for the US Government, the US utilities and the world community.[242]

One might well subscribe to Richard Garwin's[243] conclusions applicable to the US environment: "No non-weapon use has been identified for excess weapons-grade plutonium that gives it a net positive value. However, a subsidy on the order of $1 billion is a small price to pay to eliminate the special hazard of separated excess weapon plutonium."

---

[241] "How to Simplify the Plutonium Problem", Frank N. von Hippel, *Nature*, Vol. 394, 30 July 1998.

[242] "Weapons-plutonium for Electricity: A Win-Win-Win Solution", P. Goldschmidt, *Energy & Environment*, Vol. 8, nr. 1, 1997.

[243] "The 'Value' of Excess Weapon Plutonium", *op. cit.*

5. Recommendations

There is a wide consensus that both military weapons-grade plutonium stockpiles and civilian separated reactor-grade plutonium stocks should be downsized (if not eliminated) and transformed as fast as possible in a form equivalent, from a non-proliferation point of view, to the "spent fuel standard".

5.1 Military weapons-grade plutonium

- First of all, the US and Russia should agree to declare as "surplus" more than respectively 38.2 and 50 tons of weapons-grade plutonium. Retaining tens of tons of plutonium in nuclear warheads in the name of national security interests will probably not be seen as justified by a number of non-nuclear weapons states and might jeopardise efforts to enforce increased non-proliferation measures world-wide.

- The US and Russia, should pursue more vigorously the conversion of classified excess plutonium shapes into unclassified forms and subject this material to international verification measures.

- Russia has expressed a strong preference to hold its surplus weapons plutonium until it can be used as fuel in its fast breeder reactors. This will take time. In order to persuade Russia to burn the excess plutonium as MOX-fuel in its existing light water reactors (VVER type), the United States and the European Union should contribute significantly to the financing of the Russian plutonium disposal program. Nuclear weapons States, and the US Government in particular, should recognise that the cost of reducing the current threats by achieving the spent fuel standard will actually be quite modest when compared with the security benefits to be achieved.

- Licensing US light water reactors to burn MOX fuel elements should be a priority, and lead test assemblies should be fabricated in Europe, in order to gain experience, before a MOX fabrication plant becomes operational in the US. This would in no way constitute a reversal of US non proliferation policy but would on the contrary increase the US's involvement in, and impact on, international control regimes and non-proliferation technology development.

5.2 Civilian separated reactor-grade plutonium stocks

In order to accelerate the transformation of civilian separated reactor-grade plutonium stocks into spent fuel:

- Governments and safety authorities in Europe and Japan should give the highest priority to authorising and licensing additional light water reactors to recycle MOX fuel.

- In Europe, utilities willing to burn civilian plutonium from other Euratom countries (because they would find it economically advantageous) should be allowed to do so by their governments.

In so doing, governments and utilities in Japan and Europe would make a valuable contribution to the ongoing task of assuring that nuclear energy can fulfil its role in meeting growing world electricity demand in a safe, secure and proliferation-resistant manner.

Appendix: Cost Comparison of Plutonium Disposal Options (A Numerical Example)

As recalled in paragraph 4.2, the current stock of separated civilian UK plutonium amounts to more than 50 tons. The owner of most of this stock is BNFL. BNFL can either vitrify this plutonium after mixing it with HLW or sell it as MOX fuel. The purpose of this appendix is to show that under reasonable cost assumptions it can be beneficial to both BNFL and the utilities to use this separated plutonium as MOX fuel instead of vitrifying it as a waste.

(a). Let us consider a commercial price for MOX fabrication of 2000 $/kgHM. This figure is quite high and should diminish in the coming years. The *cost* to BNFL would be some 20 % less than this selling price, i.e. 1600 $/kgHM.

(b). Richard Garwin (see his chapter on plutonium disposal in this book) considers that a standard UO_2 assembly including its enriched uranium content cost 1400 $/kgHM. We will assume that in order to buy a MOX fuel element instead of a standard UO_2 element, a utility would request a discount of 30 % or some 400 $/kgHM, and would thus be willing to pay 1000 $/kgHM to BNFL for a MOX fuel assembly.

(c). The cost for BNFL to get rid of its plutonium would therefore be 1600 – 1000 = 600 $/kgHM. If its plutonium is of reactor-grade quality, i.e. with a Pu-240 content above 24 %, its MOX fuel element could contain about 7.5 % Pu (as is the case in the MOX fuel used presently in Belgian reactors). The cost to BNFL would then be: 600 $/kgHM / 7.5 % kgPu/kgHM = 8000 $/kgPu.

(d). From our estimate[244], the cost to BNFL for the storage and vitrification of its plutonium considered as a waste would amount to some 12 000 $/kgPu, i.e. 50 % more than in the case of the MOX option.

(e). This calculation shows that, provided of course the above cost estimates are correct, both BNFL and the utility would make a significant profit by following the MOX option instead of considering Pu as a waste. A great number of factors are involved in such a comparison and only a real transaction based on a commercial offer would be totally convincing.

N.B. In case weapons-grade plutonium is used in MOX fuel, its content would have to be limited to some 5 % Pu, which would make the MOX option less attractive. A remedy would be to blend this weapons-grade plutonium with "dirty" reactor-grade plutonium which would have the double advantage of degrading immediately the quality of the plutonium for weapons use and improving the economic advantage of its disposal through MOX fabrication and recycling.

---

[244] "Weapons-plutonium for Electricity: A Win-Win-Win Solution", *op. cit.*

Chapter 14 The Disposal of Plutonium

by
Richard L. Garwin*

Introduction

The first nuclear explosion on Earth, in July 1945, used about 6 kg of plutonium to produce an explosive energy release equivalent to that of 20 000 tons of high explosive (20 kilotons). Plutonium has a half-life of 24 000 years and emits essentially only alpha particles. Protected against reaction with water or air by a thin metallic coat, plutonium can be safely handled with bare hands. The absence of penetrating gamma rays makes it difficult to detect at any distance. In production reactors, natural or enriched uranium is used to sustain a neutron chain reaction by fission in the ^{235}U that makes up 0.7 % of natural uranium. The ^{238}U that constitutes the other 99.3 % cannot be fissioned with thermal neutrons, but does absorb neutrons to form ^{239}U, which decays in a couple of days to ^{239}Np and in turn to ^{239}Pu. About a gram of plutonium is made for each day of operation of a reactor with a megawatt of thermal power, so that a reactor operating at a power level of 1000 megawatts produces 1 kg of plutonium every day.

1. Weapon Plutonium

The United States has produced in this way about 100 tons of so-called weapon plutonium (W-Pu) and the Soviet Union (and now Russia) almost twice that much. The US stockpile was revealed in detail by the Department of Energy, but no similar publication is available from Russia. The first plutonium weapon was tested in July 1945 in New Mexico and used against the city of Nagasaki in August 1945. It contained about 6 kg of W-Pu in an implosion system of high explosive and gave an explosive yield of about 20 000 tons of high explosive. According to a reliable reference[245], US nuclear weapons may now have an average of 4 kg of W-Pu, although they are mostly two-stage thermonuclear weapons with the plutonium in a "primary" nuclear explosive that compresses a "secondary" explosive containing fusion fuel.

---

* Richard Garwin is Philip D. Reed Senior Fellow for Science and Technology at the Council on Foreign Relations, a Fellow Emeritus of the IBM Thomas J. Watson Research Center, and Adjunct Professor of Physics at Columbia University.

[245] Holdren J.P., Kelleher C.M., Panofsky W.K.H., Baldeschwieler J.D., Doty P.M., Flax A.H., Garwin R.L., Jones D.C., Keeny S.M., Lederberg J., May M.M., Patel C.K.N., Pollack J.D., Steinbruner J.D., Wertheim R.H. and Wiesner J.B., *Management and Disposition of Excess Weapons Plutonium*, Report of the National Academy of Sciences, Committee on International Security and Arms Control, January 1994.

Civil plutonium (or reactor-grade plutonium: R-Pu) arises in power reactors which burn enriched uranium for a longer time, in order to minimize fuel cycle costs. After yielding a heat output of some 40 000 megawatt-days per ton (MWd/ton) (3.5 EJ/kg) of fuel, containing initially some 4.4 % ^{235}U, the residual spent fuel will have a ^{235}U content of about 0.9 %, a plutonium content of about 1 %, and a fission-product content of about 4 %. In contrast to W-Pu which is typically 93 % ^{239}Pu and 6 % ^{240}Pu (of 24 000-year half-life and 6000-year half-life, respectively), R-Pu has something like 60 % of ^{239}Pu, 15 % of ^{241}Pu (also fissile with thermal neutrons) and the rest predominantly ^{240}Pu. But plutonium metal, separated from spent fuel from an ordinary power reactor, may still be a good material for making nuclear explosives. Although ^{240}Pu is not fissile with thermal neutrons, it is readily fissile with neutrons of the energy produced in fission. In fact, less ^{240}Pu is required to make an implosion-type nuclear weapon than ^{235}U.

It has now been widely and authoritatively established that reactor plutonium can be used to make powerful and reliable nuclear weapons.[246] The simple substitution of R-Pu for W-Pu (with a slight increase in amount to make up for the larger critical mass of the ^{240}Pu component) would require some attention to managing the greater heat evolution from R-Pu than from W-Pu, and, if no other changes were made in the Nagasaki weapon, would produce an explosive release of 1000-2000 tons of high explosive (never less). But anyone capable of designing a nuclear weapon could use a more advanced design to produce essentially the same yield with R-Pu as with W-Pu, including high-performance two-stage thermonuclear weapons.

The amount of reactor-grade plutonium already separated from spent reactor fuel is vast, as reported to the International Atomic Energy Agency (IAEA): see table 1.[247] Thus, the separated stocks of R-Pu total about 250 tons, but about 35 ton additional are separated each year - largely by the *Compagnie des Matières Nucléaires* (COGEMA), in France, and British Nuclear Fuel Limited (BNFL), in Britain at Sellafield.

2. Building On Other Chapters In This Book

In order to avoid redoing what has already been done, in this book, by Pierre Goldschmidt, in his chapters on "Spent Fuel Management" and "The Disposal of Separated Plutonium Stocks", I will refer to his text. There are, however, some points which should be clarified. As a member of the Committee on International Security and Arms Control (CISAC) of the US National Academy of Sciences (NAS), I contributed to both the 1994 volume on "Management and Disposition of Excess Weapons Plutonium", and the 1995 (larger) volume on the same subject, entitled "Reactor-Related Options". It is quite true that the only two options that CISAC judged practical and competitive were disposal of W-Pu vitrified with high-level fission products and burning W-Pu as Mixed Oxide fuel (MOX) in reactors of existing types, and disposing of the spent fuel into

[246] *Management and Disposition of Excess Weapons Plutonium, op. cit.*, pp. 32-33.

[247] André Jaumotte and Alain Michel, Table 1 in "Plea for a Civilized Use of Plutonium", 11 November 1998, paper for the Eleventh Amaldi Conference, Moscow, 18-20 November 1998.

mined geologic repositories without reprocessing. This was not a prejudice, but a CISAC judgment, based on the analysis of several other proposals current at the time, e.g. shooting plutonium into the Sun, dispersal in the water of the oceans, sub-seabed disposal, etc., which analyses are reported in Appendix C of the 1994 report.

	Held in the country itself (kg)	Held in other countries (kg)
Japan	5 000	15 100
Germany	4 900 [(a)]	-
Belgium	2 700	-
Switzerland	100 [(b)]	less than 50
France	65 400 [(c)]	200
USA	45 000 [(d)]	0
China	0	0
UK	54 800 [(e)]	900
Russia	28 133	-

(a) 6000 kg as of 31/12/97.
(b) 700 kg as of 31/12/97.
(c) Includes 30 000 kg belonging to foreign countries.
(d) This is the amount of civil plutonium; 52 500 kg previously military plutonium, declared "in excess" recently, is probably not included in the above.
(e) Includes 3 800 kg belonging to foreign countries.

Tab.1. Holdings of civil un-irradiated plutonium:
quantities in kilograms, holdings in the country as of 31 December 1996.
Source: http://www.iaea.or.at/worldatom/infcircs/98index.html).

Goldschmidt writes:

"Spent nuclear fuel is therefore considered by the US National Academy of Sciences to be the standard or paradigm of an effective proliferation-resistant form in which to confine plutonium. This level of proliferation resistance has become known as the "spent fuel standard" and has been adopted by the United States Department of Energy (US DOE) as an objective for the form into which one should transform surplus ex-military plutonium, before its disposal in an appropriate geologic formation."

In fact, CISAC did not judge this as proliferation-proof: our considerations were somewhat more complex. At the time, there were many proposals for the essentially complete elimination of excess W-Pu by shooting it into the Sun, by repeated burning in reactors until nothing was left, and by other extreme measures. We proposed the "spent fuel standard" with full knowledge that there is a much larger amount of reactor-grade plutonium (R-Pu) than W-Pu, in the form of spent fuel. We judged that if one could convert the W-Pu into a form that was approximately as difficult to use for fabricating a nuclear weapon as un-reprocessed spent fuel, then this was likely to be the optimum. To go to extreme measures to totally eliminate excess W-Pu would do less for hampering proliferation than would bringing the W-Pu to the spent fuel standard, and then using any additional money that would have been spent on elimination of W-Pu to attack the entire plutonium problem. So the spent fuel standard is not an absolute judgment that spent fuel itself is adequately proliferation resistant; we recognize explicitly that spent fuel represents a problem still to be managed.

Goldschmidt comments about the difficulties the United States has had in moving rapidly to handle its excess plutonium stockpiles, and challenges its commitment to solving this problem, "The more so when, at the same time, the US DOE is considering spending more than $10 billion to construct and operate an accelerator for the production of tritium, a radioactive gas used to boost nuclear weapons' explosive power". I have long been involved in studies of tritium supply, and have argued strongly that it would be far less costly for the US to acquire and maintain a tritium stockpile through purchase on the open commercial market. If DOE is unable to buy tritium abroad, it should choose the least-cost domestic option: by using existing reactors producing electric power, one avoids spending $6 billion for a particle accelerator or $2.3 billion for a reactor to do the same job. Late 1998, the US DOE announced that the US would indeed use irradiation services of lithium in commercial electricity-generating reactors, thus eliciting my personal support.[248]

Goldschmidt also quotes the Center for Strategic and International Studies (CSIS) 1998 report that states that "First, as a matter of high priority, the two countries should start converting their excess stocks that are now in classified shapes (such as plutonium pits) to unclassified forms". Indeed, it would be desirable to do that, and it is an essential step on the way to disposal of excess W-Pu, but one can ensure that US pits can never be used in a nuclear weapon by a simpler procedure, and one that can be carried out on intact nuclear weapons. This approach ("pit stuffing") has been explored by Matthew Bunn and myself, although whether Russian nuclear weapons are amenable to this procedure is not clear to us.[249]

Goldschmidt provides a great deal of useful information and quotations from reasonably authoritative studies but in quoting the February 1998 Royal Society "Management of Separated Plutonium" I believe that a mis-impression may be conveyed: "But given that reprocessing has been carried out and the plutonium is available, then the extra cost of fabricating MOX fuel (as compared with enriched uranium fuel) might be justified in view of the savings in uranium and enrichment costs". But the Royal Society says "might". It would be very useful for BNFL in particular to understand whether it has 50 ton of valuable material or 50 ton of liability on its hands.

---

[248] See Richard L. Garwin, letter to the editor of the New York Times (unpublished), 30 November 1998, and letter to the editor of the New York Times, 23 December 1998, published 26 December 1998.

[249] M. Bunn, commented by R.L. Garwin, "Pit-Stuffing: How to Disable Thousands of Warheads and Easily Verify Their Dismantlement", *FAS Public Interest Report*, Vol. 51, No. 2, March/April 1998. See http:// www.fas.org/rlg.

Goldschmidt then states:

"Some experts have attempted to demonstrate that recycling separated plutonium as MOX-fuel in Europe is not economically viable. These calculations usually compare all the additional costs associated with the fabrication, licensing and use of a fresh MOX-fuel element, to the cost of a fresh standard uranium fuel element, and conclude that the value of plutonium is negative. This approach fails to take into account the fact that the already separated plutonium, if not recycled in MOX-fuel to produce electricity and reach the spent fuel standard, has to be stored, immobilized and disposed of as waste and could therefore have a significantly greater negative value."

It may indeed be true that the excess cost of fabrication into MOX and burning is less than that of direct disposal of plutonium, which clearly costs money. But this is not the same as saying that the plutonium has a positive value. The reactor operator has no incentive to buy MOX fuel under these circumstances, and it is the holder of the separated plutonium oxide who will have to provide the funds to drive this process. It is possible that national governments would be willing to provide a subsidy. It is almost certain that the holders of excess separated plutonium will request it.

Goldschmidt then writes:

"The question is then to find out whether the negative value assigned to excess separated plutonium by the UK owner is sufficient to make the deal attractive to both the plutonium owner and the utility."

I do not see how a negative value would make the deal attractive to both partners in the activity. BNFL has separated this plutonium and owns it. Presumably, it has the responsibility to dispose of it.

3. Prescription

The prescription varies as between US excess weapon plutonium, Russian excess weapon plutonium, civil plutonium already separated from spent fuel, and civil plutonium in highly radioactive spent fuel.

3.1 Excess US weapon plutonium

In the US bureaucracy, every step seems agonizingly slow. The government must analyze every reasonable option (and many unreasonable options) and in the process announce what it is doing, make time for comment, file preliminary and final environmental impact statements, and the like. This procedure helps avoid mistakes, but it greatly increases cost and delay, and provides opportunities for groups that disagree with an approach to delay it even if they cannot prevent its implementation.

After plutonium is removed from the pits[250], an automated tool divides it into two hemi-shells. A hemi-shell is held with the plutonium side downward in a hydrogen atmosphere, causing the plutonium to hydride and to fall off. Heating the plutonium hydride liberates the hydrogen to scavenge more plutonium in this way. The plutonium hydride can then be oxidized in a single step or in a two-step process to yield plutonium oxide that is not yet suitable for making MOX. The high-temperature delta-phase plutonium in US nuclear weapons has been stabilized over the environmental temperature range by the addition of on the order of 1 % of gallium. This gallium must be removed if it is not to cause problems by chemical reaction with the sheath on the fuel rod in a reactor. The US regards the plutonium as unclassified once it is transformed into a metal ingot or oxide in this fashion. The US will either need to build a MOX plant or arrange for foreign fabrication of MOX to the extent that it will dispose of its excess W-Pu by burning it in US reactors.

As for the vitrification option: that will be carried out at the Savannah River Plant (SRP), where some 8000 canisters containing about a ton each of vitrified high level waste are to be produced over the next decade or so. In principle, plutonium oxide or plutonium solution could be added to the solution of fission products that go to the "calcination" unit for drying and to the "smelter", for melting with borosilicate glass powder. A level of about 2 % W-Pu in this composite would correspond to about 20 kg of W-Pu per canister, and 2500 canisters (1/3 of the total) would suffice to dispose of the 50 tons of W-Pu. It was early established that the glass would dissolve at least 7 % Pu in the form of oxide, so little problem was envisaged with this approach.

However, SRP has chosen a very large smelter (in contrast with the small smelter used routinely by COGEMA) to dispose of the vitrified fission product waste resulting from commercial reprocessing. For the large smelter it is difficult to absolutely guarantee that the plutonium does not congregate in a portion of the smelter rather than dissolving uniformly in the glass. This would lead to criticality and a small explosion that would locally disperse radioactivity.

So the baseline program (with development being done at the Lawrence Livermore National Laboratory) involves incorporating the plutonium in ceramic disks, each contained in a stainless steel shell. Shells are to be mounted on a structure erected within the stainless steel can, ready to receive the molten fission product glass. While there are differences in the approach to separating plutonium from highly radioactive glass on the one hand, or from ceramic Light Water Reactor (LWR) fuel on the other, it is judged that the difficulties are roughly comparable, so that either form meets the spent fuel standard. However, the recovery of W-Pu from otherwise non-radioactive ceramic in these small cans would be far simpler, and the acceptability of can-in-canister as meeting the spent fuel standard was made the subject of inquiry of a short study begun by the US National Research Council in November 1998.

---

[250] In principle by a "dry" process that involves disassembly in a nuclear weapon, removing the high explosive from the pit, and then holding the pit in a jig.

There is clearly the necessity for interim storage, of the surplus nuclear weapons, of the plutonium-bearing pits removed from the weapons, and perhaps at other stages of this process. And it would be useful if weapons, once declared excess, were irrevocably transferred to the civil inventory, and if even before dismantling could take place, they were demilitarized permanently by stuffing the hollow pit with metal wire or fragments that cannot be removed without destroying the pit and re-fabricating it.

3.2 Russian excess weapon plutonium

Although a "lab-to-lab program" between US DOE laboratories and Russian establishments has in the last few years substantially improved the security and accounting for Russian weapon plutonium and highly enriched uranium, it is still not up to Western standards. And there is no guarantee that Western standards are good enough, either. As indicated by Goldschmidt, Russia has been very resistant to the suggestion that they might dispose of some excess weapon plutonium together with vitrified fission product waste. In fact, enormous amounts of such high level wastes were disposed of over the years (extending certainly to 1994) by direct injection into disposal wells drilled in the ground. As indicated by Goldschmidt, the process of pit disassembly, MOX production, and MOX fuel fabrication is lagging, because of a lack of necessary decision and a lack of funds to carry out this process as a high priority.

3.3 Disposal of separated civil plutonium

As indicated, civil plutonium can be used to make powerful nuclear weapons, and it is in general not guarded with that hazard in mind. In addition to being under IAEA standards of accountancy and safeguards, civil plutonium should be afforded adequate physical protection, and the developed world should have a judgment on and a stake in its security.

3.4 Spent fuel containing plutonium

With more than 400 nuclear power reactors the world over, and with spent fuel being largely stored at reactors, it is possible that this highly radioactive material could be diverted and processed by groups or nations to yield plutonium that could be used for making nuclear weapons. In the United States, the US Department of Energy was supposed to begin accepting from the reactor operators (beginning January 1998) the spent fuel for which the operators have been paying one-tenth of a cent per kWh for just this purpose. The fuel was to have been shipped in transport and storage casks for an additional cooling period, and then placed in the Yucca Mountain mined geologic repository. Political and technical problems have delayed the opening of Yucca Mountain, reportedly to at least the year 2015, and there is now a dispute between the

Department of Energy and the reactor operators over what should be done. I believe that DOE should step up to its responsibility and store the fuel above ground until the facility is open. But there is no local economic driving force for this approach. And many officials and forces of the population want to keep this nuclear waste out of their backyards.

4. Competitive, Commercial, Mined Geological Repositories

I have long advocated that some countries should take the lead in changing their laws so as to allow them to dispose of IAEA-approved forms of spent fuel and vitrified fission products, for a fee.[251] This would be a commercial mined geologic repository. When others see that there is money to be made in this approach, they, too, will do similar things, and now there will be competition. Tom Cochran and Chris Paine propose an ingenious scheme in which the driving force (the money) for the storage and disposal of excess Russian weapon plutonium comes from fees paid to an organization to accept and dispose of commercial spent fuel.[252] They assume that the reactor operators would pay a fee which is comparable to that paid now for spent-fuel storage, reprocessing, and high-level waste disposal (about $1.2 million per ton).

The organization would contract with the Russian Ministry of Atomic Energy (MINATOM) to store the foreign power reactor spent fuel for say 30 years at Krasnoyarsk-26 or another suitable site. This would require amendment of Russian Land Use Law Number 53, storing the spent fuel above ground in dual purpose, dry casks. At the same time, the organization would lease from MINATOM for a specified period (say 20-30 years) up to 50 tons of excess weapon plutonium. They suggest that the acquisition of plutonium be paced by the transfer of spent fuel, according to the amount of plutonium in the spent fuel (although this has no fundamental relevance). The "organization" mentioned here would be the Non-Proliferation Trust, and the proposal has been approved by the Trust and its shareholders, which include the MINATOM Development Trust.

5. The Highly Enriched Uranium Problem

Russia has probably some 1200 tons of highly enriched uranium, of which the United States has agreed to buy 500 tons over 20 years for about $12 billion. Many problems beset this deal, including the privatization of the United States Enrichment Corporation (USEC) in July 1998. USEC is the executive agent for this purchase, but USEC also has a duty to its stockholders to maximize its profit. The "enrichment services" component of the enriched uranium from Russia exceeds the cost of enrichment

[251] See for example my paper at the 29th Annual Japan Atomic Industrial Forum Conference in Nagoya, at http://www.fas.org/rlg.

[252] T.B. Cochran and Ch.E. Paine, *Proposal for Augmenting Funding for the Disposition of Russian Excess Plutonium*, Natural Resources Defense Council, 18 November 1998.

performed in the United States, so USEC has incentives to slow the deliveries, encourage MINATOM to back out of the deal, etc. Furthermore, although it would be useful to blend down the typically 93 % highly enriched uranium (HEU) to 19.9 % (which no longer is categorized as highly enriched uranium by IAEA), blending Russian HEU with natural uranium, or with depleted uranium, produces a material with about 1.5 % U^{234}/U^{235} ratio, rather than the 1 % standard that has been established for uranium for fabricating fuel for light-water reactors. This arbitrary limit increases the cost of disposal of HEU, but it also greatly reduces the rate of which the blending could be achieved. The U^{234}/U^{235} standard should be relaxed, while guaranteeing a similar level of protection for the workers.

One might obtain from the G-7 an offer to pay $24,000/kg for HEU delivered in Russia for this program within 3 months, $22,000/kg for material delivered in 4 to 6 months, and $20,000/kg for material delivered after that time. The material would be blended down to 19.9 % so as to maintain maximum flexibility for its later use in starting breeder reactors, in various types of LWRs, etc. Various treatments of the money paid are possible, and Russia might retrieve HEU from this stock by refunding the purchase price.

6. The Long-Term Future

As emphasized in my chapter (11) on reprocessing, unless the cost and performance of breeder reactors is entirely comparable with that of LWRs, even an all-nuclear future can be maintained for thousands of years using uranium from seawater at costs of $100/kg or $300/kg. Vast amounts of plutonium will arise each year in spent fuel from reactors - 200 tons per terawatt-year of nuclear power - and even this current flow could supply some 40 000 nuclear weapons per year. If there is reprocessing, the flow of separated plutonium will be some fraction of this amount. If societal disruption causes the industry to shut down, this plutonium will need to be protected in order not to be available for use in nuclear weaponry.

The existence of several or even 20 mined geologic repositories for spent fuel would not add significantly to the problem, even though after 500 years the gamma radiation from fission products would no longer provide a significant barrier to impede the process of extraction of plutonium from the spent fuel. It is essential to understand that the danger from nuclear weapons arises from the first few, or the first hundred, or the first 10 000; the fact that there is a repository that contains 80 000 tons of spent fuel and hence 800 tons of plutonium (as will be the case with Yucca Mountain) poses no greater hazard of proliferation than if it contained one tenth that amount and could provide plutonium for (only) 160 000 weapons.

If mankind is incredibly lucky and so successful that it survives to begin to exhaust the uranium in seawater, and if we have not been successful in mastering the economic generation of electrical power from fusion or from sunlight, we will still be able to use the breeder reactor to burn essentially all of the natural uranium, the depleted uranium from which was made the low-enriched uranium, and also the plutonium in the spent fuel that will have been buried in the repositories. But the energy value of the plutonium will be only 1 % that of the uranium in the spent fuel, and plutonium will not be necessary to begin the operation of a vast population of fast-neutron breeder reactors. It would suffice to extract some 11 tons of HEU from 2000 tons of natural uranium to begin the operation of a breeder, which could then continue on recycled uranium and plutonium.[253] To fuel 10 000 breeder reactors would require a mere 20 million tons of uranium (of the 4 billion tons in the oceans of the world); we will not have forgotten how to enrich uranium, and even the current costs and techniques of enrichment are entirely adequate to this task.

The priorities for the next decades are:

- In the next year for the G-7 to provide a plan and financing for the rapid protection and future of excess Russian weapon uranium and plutonium.
- For the nations of the world and the nuclear industry to modify the laws so as to permit the acquisition and disposal of IAEA-approved spent fuel and vitrified fission products.
- For nations and commercial interests to create competitive, mined geologic repositories for IAEA-approved storage forms, under IAEA safeguards and international protection.
- For nations with an interest in the energy future to support research on the acquisition of uranium from seawater in order to determine the cost of this great energy resource and thus to guide or to delay the investment in breeder reactors and the reprocessing of spent commercial fuel.
- For nations to explore concepts such as the lead-cooled breeder reactor with the understanding that it would be deployed only if it were cheaper to build than the existing commercial power plants; and similarly for other concepts such as the gas-turbine, modular high-temperature gas-cooled reactor.
- For nations and the public to fully support and perfect the IAEA-based set of safeguards; and to expand the protection against diversion of fissionable materials to build nuclear weapons, including by the provision of appropriate IAEA safeguards over civilian activities in nuclear-weapon states.

---

[253] "The Role of the Breeder Reactor", chapter of the book *Nuclear Energy and Nuclear Weapon Proliferation*, ed. F. Barnaby *et al.*, Taylor and Francis, London, 1979, pp. 141-153.

Chapter 15 Fast Neutron and Accelerator-Driven Reactors

By Georges A. Vendryes*

1. The Prospects of Nuclear Energy

Figure 1 shows how the world population evolved over the last few centuries, and how it is expected to evolve till the end of the next one. In 1900 it was less than two billion people and it will exceed six billion in 2000. Possibly it will stabilise during the second half of the twenty-first century, somewhere around ten billion under conservative assumptions. The demographic explosion, which took place in the course of the past hundred years, is a unique phenomenon in the history of mankind. (Be it said incidentally, its magnitude is such that the number of people living on earth today exceeds 10 % of the total number of those who have lived since the human species emerged among the primates, several million years ago). It is also the greatest challenge that faces us.

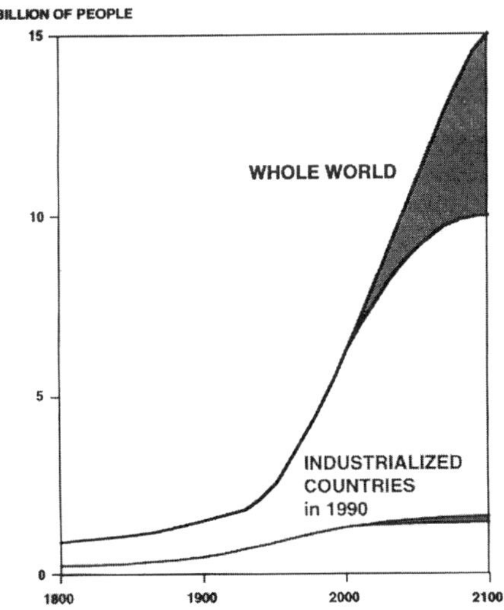

Fig. 1. World Population Growth.

What makes such a large and rapid increase dramatic is the growing gap between the standard of living of affluent countries, whose population has already reached a more or less stable level, and that of poor countries, whose population continues to expand.

* Georges Vendryes is a former Director of the Industrial Nuclear Applications Division of the French Atomic Energy Commission (CEA).

That such a distressing situation endures is unbearable. It is not only a moral issue, but also the selfish interest of all, both rich and poor. I do not see how our small planet could enjoy lasting harmony and peace so long as shocking disparities exist between its inhabitants, as far as food supply, medical care, education, etc. are concerned.

This applies particularly to energy needs. The total annual consumption of energy world-wide, which has increased more than six times over the past one hundred years, (see figure 2) is at present equivalent to 8 billion metric tons of oil. But its use is extremely unequal among the various countries, as shown for example on figure 3. Whereas an average citizen of Canada or Scandinavia disposes of the equivalent of almost 8 metric ton of oil each year, a Nepalese or an Ethiopian can hardly afford more than a few tens of kilograms.

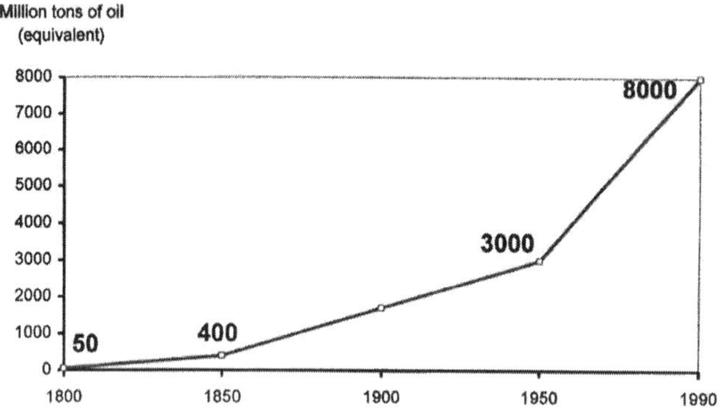

Fig. 2. Yearly World Energy Consumption.

Yet there is no problem of supply. Solar radiation pouring down continuously on earth is a heat source of low entropy which exceeds, by many orders of magnitude, the present and future energy needs of mankind, and which is practically inexhaustible. It does not mean that man is entitled to squander it, in so far as a thoughtless use of the available energy resources might disturb in an irreversible way the fragile equilibria on which the unique biosphere of our planet is balanced.

Nobody would deny, to citizens of countries which are still now developing, the right to enjoy as soon as possible a standard of living comparable with ours. Thanks to the further remarkable progress to be expected from science and technology, the energy intensity - i.e. the ratio of energy consumption to gross domestic product - is likely to be in the future significantly lower than it is in most rich countries at present. Even so, a considerable increase in the world energy consumption will take place over the next century. It is not only unavoidable; it is desirable and needful, with the proviso that appropriate measures be taken to protect the environment.

The challenge to be met is so serious and difficult that no available source of energy should be discarded. In my opinion they complement instead of oppose each other. It is not intended to enter here into any detailed comparison of the possibilities, merits and drawbacks of the few methods we know today to generate large amounts of electricity at low cost and in a safe way. Over a time scale limited to the next century, there are only three of them, namely combustion of fossil fuels, conversion of solar radiation and nuclear fission.

No mention of nuclear fusion has yet been made. This omission was not inadvertent. I am firmly convinced, for a series of reasons, that the possibility to produce any significant amount of electricity in a machine where a thermonuclear plasma of deuterium and tritium is confined in a magnetic bottle is out of the question for the next hundred years.

1992 CONSUMPTION OF COMMERCIAL PRIMARY ENERGY
(tons of oil equivalent per head)

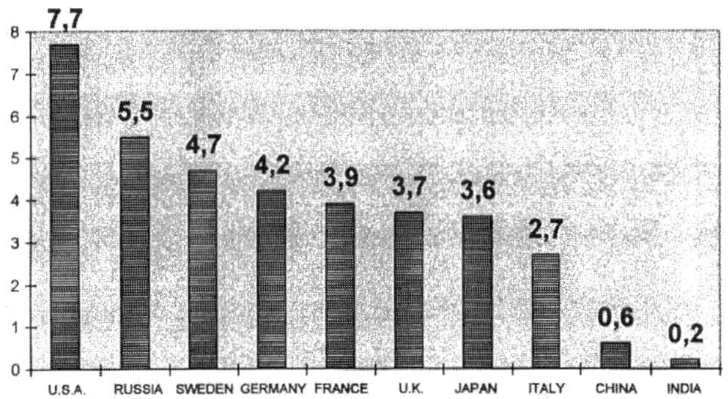

Fig. 3. Consumption of Commercial Primary Energy in 1992.

The use of wind, tides or waves as power sources can be marginal only, and the same is true for geothermal energy. As far as hydropower is concerned, a large fraction of the suitable sites world-wide have already been equipped.

I am very much in favour of developing all possible ways to use the huge energy flux of the solar radiation. It is the only known source which we can tap forever, and it is expected to take a growing part in our energy mix, by direct conversion (photocells) or through indirect means (biomass). The main drawback of incident solar radiation is its low energy density. It is inherently not well suited to the installation of large and cheap power generation plants, at a time there is a world-wide trend towards concentration of people into bigger and bigger cities, whether this should be deplored or not.

The true choice for the near future is between burning fossil or nuclear fuels. The present low market price of gas makes it now the preferred option for many customers, and everybody is now rushing into burning it, for various purposes including electricity generation. For some important applications, like the transport sector, there is no substitute for oil or liquefied gas. Many poor countries do not have the financial means, nor the industrial infrastructures, nor particularly the ability to train qualified personnel, to design, build and operate installations like nuclear power plants, which are and will remain high-tech products. For those reasons, and many others, one can be sure that coal, oil and gas will continue to occupy a dominant part in the energy mix of the forthcoming decades, during which a significant fraction of the fossil fuels formed over the past hundreds of millions of years will progressively tend to become extinct.

At the rates of consumption which can be expected, the production by man of carbon dioxide and of other gases which display a greenhouse effect or are noxious to health is soon likely to exceed what seems to be acceptable. Their real influence on global climatic changes is still arguable. Even so it appears wise to take henceforth conservative measures, as irreversible disturbances could be set in motion, which it would be to late to counteract once they are established beyond doubt. The best available means to cope with this problem is to develop on a large scale the use of nuclear energy. This was clearly perceived, albeit acknowledged only by lip service, at the recent Buenos Aires conference. On that score the example of France is particularly relevant, as shown in figures 4 and 5.

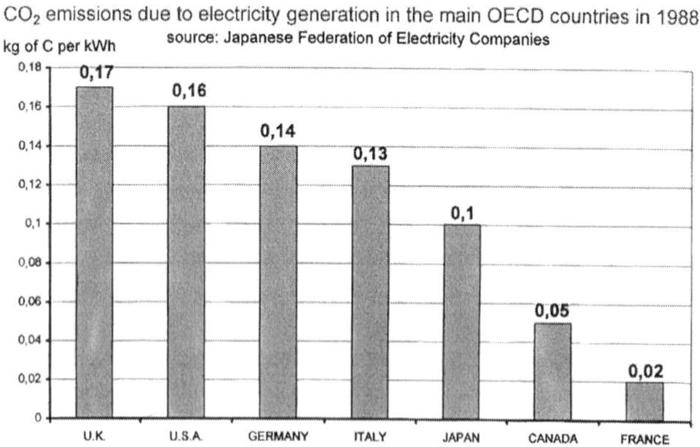

Fig. 4. CO_2 Emissions due to Electricity Generation in the main OECD Countries in 1988.

A strong opposition to the use of nuclear energy, which started in the United States about thirty years ago, has now developed in many other places, especially in the western world. Its rejection is based on a number of fears: radioactivity, impervious to our senses, is felt mysterious and alarming; nuclear plants are deemed liable to big accidents; a link is established with the dreadful nuclear weapons; the possibility of the management of long life nuclear wastes is called in question.

All these matters must be dealt with in a very serious and responsible way, but without yielding to passion and by keeping in mind that every course of action presents its specific disadvantages and difficulties. No human activity is wholly exempt from risks. Comparisons between those of various energy systems have been repeatedly made. Their conclusions are not to the disadvantage of nuclear energy.

Satisfactory answers can undoubtedly be given to the above mentioned questions, indicating that nuclear energy from fission can be deployed for the benefit of mankind, in a safe way, benign to the environment, providing that a series of measures and precautions, which are well identified and which improve continuously, are taken.

Unfortunately, events occur everyday that show how difficult it is to treat those topics dispassionately. Indeed, the whole issue has become the preferred area for irrational trends to surge. An article published mid-November 1998 in a French magazine by Georges Charpak, recalling the merits of nuclear energy and exposing the fallacies of those who attack it is a recent example.[254] There are constant examples of public opinion being led astray by biased information or false data. Too often some media are prone to add fuel to the fire in order to excite people through thrilling, yet unfounded news. One can find extreme cases of violent antinuclear actions conducted by terrorist groups.

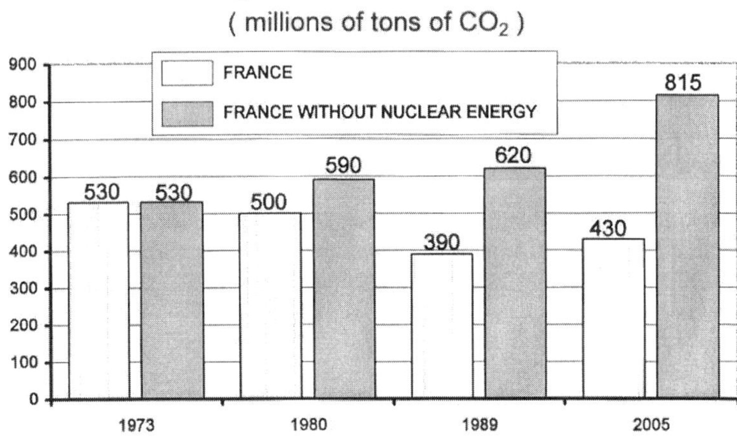

Fig. 5. CO_2 Emissions in France.

Such a state of things can not only jeopardise the future prospects of nuclear energy. It also entails potential dangers, which go much beyond it and which raise basic problems for the working of our societies. Is it acceptable to make shift with the current situation, where some find a flourishing stock-in-trade, simply by lamenting it? All initiatives taken to discuss in a cool and rational way the role to be played by nuclear energy must be fostered. The Pugwash movement can surely be of great help in that respect.

[254] Georges Charpak, "La supériorité du nucléaire en termes de sécurité est écrasante", *Figaro Magazine*, 14 November 1998.

2. Fast Neutron Reactors

In what follows, it is assumed that reason will carry the day, that the pressure of hard facts will finally prevail over prejudices and phantasms, and that nuclear energy will again be recognised as a major component of the energy mix in the future.

As soon as a renewal of nuclear energy will take place on a large enough scale, fast neutron reactors will again come to the forefront, because of their remarkable and unique capability of breeding nuclear fuel. It is well known that present day nuclear power plants equipped with slow neutron reactors can release only about one hundredth of the fission energy contained in natural uranium, whereas the whole of it could in principle be made available by means of fast neutron ones. Through the use of the latter, the energetic potential of uranium deposits (and of those of thorium as well) will be increased by a huge factor. Not only a hundred times but much more, because the added value granted to uranium thanks to breeding makes it economically justifiable to exploit very low grade uranium resources, the use of which would remain unwarranted if only one hundredth part of them could be turned to account.

For this reason, fast neutron reactors represent, so to say, the *nec plus ultra* of nuclear energy, the symbol and the pledge of its perennial use. Antinuclear movements were not mistaken about it. There, and nowhere else, is to be found the ultimate cause of their steadfast hostility to breeder reactors.

Two past instances of the relevance of fast neutron reactors in that respect are worth recalling. Right after the Second World War, when the American Atomic Energy Commission (AEC) was starting to investigate how to promote the peaceful applications of nuclear power, one of the main problems was the apparent shortage of natural uranium. The only large deposits of uranium ore, which were well identified at that time in the western world, were those of Belgian Congo, but most of them were earmarked for military purposes. Under such circumstances, no wonder that a top priority was assigned to fast neutron reactors in the early programs of the AEC, so much so, that the first reactor in the world to demonstrate the possibility to generate electricity was the experimental breeder reactor called EBR 1, in 1951.

Another example occurred in the wake of the 1973 and 1979 oil crises, which emphasised how much the energy needs of industrialised countries were dependent on oil supplies from the Middle-East and how their economy was vulnerable for that reason. As a consequence, large nuclear energy programs were launched in several countries. As a long-term precautionary measure against a possible shortage of natural uranium, or even before this, against an increase in its market price, a prominent part was reserved for fast neutron breeder reactors. In this context several European utilities joined forces to develop large prototype fast neutron breeders, starting with *Superphénix*. A few years later Japan extended its own program in that field by deciding to build the Monju demonstration reactor.

To be sure, the present situation is quite different. There is plenty of low price fossil fuel available on the market and no threat of energy shortage is looming ahead. In a fast changing world the behaviour of people and the policy of governments are more and more conducted by short-term motives. The high capital investments of nuclear power plants, providing a financial return on the long run, are a significant drawback in comparison with gas turbines, which can be built cheaply and quickly. Such circumstances are propitious to systematic campaigns for disparaging nuclear energy.

But there are reasons to think that the present situation is a temporary one and that the growing energy needs of the next decades can best be met with an important contribution of nuclear energy, which will imply sooner or later the use of fast neutron breeder reactors. Countries like Japan and Russia are eager to pursue their research and development programs in that field, and others, like China or South Korea, are actively preparing to enter it. Those who have already acquired an expertise in that area should be well inspired to keep it, and to maintain open an option which is likely to be needed later on. In that regard, I consider as fully irresponsible the decision taken by the present French government to shut-down *Superphénix*, on temporary electoral grounds.

In addition to their unique capability to breed nuclear fuel, fast neutron reactors can also be made the most efficient burners of fissile nuclei. At first sight it might seem paradoxical to claim they are both the best producers and the best consumers of nuclear fuel, but there is nothing contradictory there. These two characteristics are just two sides of the same picture. Both arise from basic data of nuclear physics and result from the larger number of neutrons which are available when fission chain reactions are kept going by fast neutrons.

Take for example the case of plutonium, where three neutrons are released by each fission on an average. One of them is required to produce the next fission, but the two others may be used, within some limits, for various purposes. As the fraction of neutrons lost in the reactor through a series of unwanted - yet unavoidable - nuclear reactions can be kept to a minimum when the neutrons are fast enough, there are in that case greater opportunities to use them profitably. One may design the reactor core according to the object one has in mind, and direct in various ways the neutrons in excess, either for producing a maximum of new fissile nuclei through their capture by ^{238}U or ^{232}Th, or for consuming actinides by means of fission reactions, or for destroying by transmutation a number of unwanted nuclei.

At the present juncture, there is no lack of cheap uranium on the world market, and consequently no need for breeding. At the same time plutonium produced by existing nuclear reactors piles up. The total quantity of plutonium contained in the spent fuel removed from the 450 or so nuclear power plants operating in the world goes up as shown in figure 6. It increases by about 80 metric tons each year and it will reach 1600 metric tons in 2000 (of which 1100 metric tons comprise fissile plutonium). In addition are produced minor actinides (mainly neptunium, americium and curium), the amount of which represents roughly 10 % of that of plutonium.

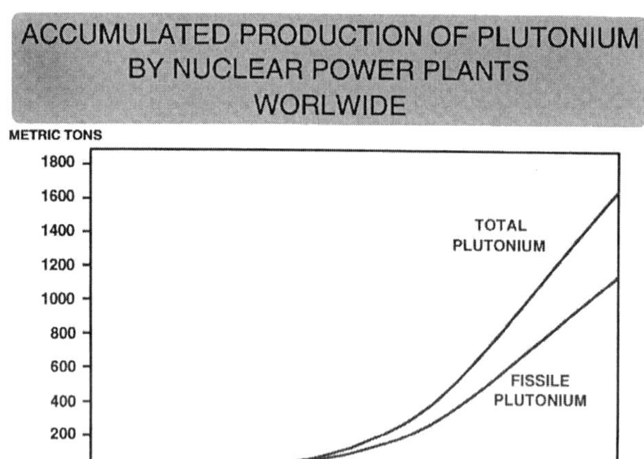

Fig. 6. Accumulated Production of Plutonium by Nuclear Power Plants World-wide.

The accumulation of plutonium and of other actinides, either within the spent fuel assemblies or separated from fission products after chemical reprocessing, is regarded by many as a problem, so long as there is no clear-cut use for them. The fact that most actinides decay with very long lifetimes is claimed to make difficult our getting rid of them by way of definitive storage in a deep geologic repository.

I think that such a concern is overdone. Spectacular progress has been made on the design of glass matrices able to retain minor actinides over periods greater than 10^4 years and in the understanding of their behaviour in specific geologic formations, in case they would be released there in a more distant future. If such results are confirmed by further scientific investigations, disposal of high level wastes containing minor actinides in a suitable deep geologic repository will appear as a perfectly acceptable solution, consistent with the stringent responsibilities we have - in this regard as in many others - towards future generations.

Concerning plutonium, far from being viewed as a cumbersome waste, it represents in my opinion a valuable energy resource, which would be foolish to ignore. The best, if not the only reasonable use one can make of it is to generate electricity by burning it in a nuclear reactor. If wanted, one can aim at burning minor actinides as well.

The near-term availability for civilian purposes of fissile materials from nuclear weapons is a highly welcome addition to the picture. The conclusion of agreements between the United States and the countries of the former USSR to progressively dismantle a part of their huge arsenals of nuclear warheads is a historical event of major importance. The total quantity of plutonium, which they contain, is generally estimated at 200 metric tons or somewhat more. Military grade plutonium is an excellent material for manufacturing fuel to be burned in nuclear power plants, which are less demanding than weapons as far as the isotopic composition of plutonium is concerned.

Fissioning and transmuting actinides is carried out most efficiently in a fast neutron reactor, for several reasons. Some fission reactions can only take place with neutrons whose energy exceeds a certain minimum value. Besides we have already mentioned that the number of neutrons available, notably for transmutation purposes, is greater in a fast neutron reactor. Last but not least, in a slow neutron reactor the quality of plutonium as a fuel diminishes progressively with its irradiation, as a consequence of the progressive formation of isotopes, which act like poisons, in a slow neutron spectrum. Roughly speaking, fast neutron reactors are largely indifferent to the isotopic composition of the actinides they are supplied with, and they can devour all of them.

To be more specific, let me give you some information about the situation in France. The 57 French pressurised water reactors (PWRs) now licensed to operate would produce a total of roughly 14 metric tons of plutonium each year, if they would all use a standard fuel made of slightly enriched uranium. So far, 20 of them (units of 900 MWe only) have been authorised to have one third of their core loaded with the so-called Mixed Oxide (MOX) fuel, made of a mixture of 6 % plutonium and 94 % of depleted uranium. In such a case the quantities of fresh plutonium produced, and that of plutonium destroyed by fission during the same time, are about the same, so that there is no net increase in the amount of plutonium present in the whole spent fuel in comparison to what it was to start with. When all 20 units are operating under those conditions, the yearly production of plutonium by French PWRs will be reduced to something like 10 metric tons.

In the future, it appears possible to slow down further the rate of plutonium production, and even to reduce the accumulated stocks, by designing advanced PWRs which would accommodate MOX fuel in their whole core. Yet there are drastic limitations along this line. In France there exists no provision at present for more than one plutonium recycling. Thus, however interesting it is, consumption of plutonium as MOX fuel in a PWR can only be viewed as a partial solution.

The true alternative for the long run, the only way to destroy efficiently and usefully actinides, and plutonium in the first place, is to burn them in fast neutron reactors specially designed for that purpose. There will be no blanket around their core to give rise to new fissile materials. Their core will be designed so that the amount of plutonium destroyed by fission in excess of that produced by neutron captures in ^{238}U nuclei would be as large as feasible. This objective will be reached by using a fuel where the ratio of depleted uranium to plutonium is as low as possible. Figure 7 shows how the net consumption of plutonium depends on that ratio. If the fuel could be made of plutonium alone, with no depleted uranium at all, the reactor would burn 110 kg of plutonium for each TWh (billion of kWh) of electricity generated. There could be technical reasons to keep some uranium within the fuel elements. Even so, it is seen that in a fuel made half of plutonium, and half of depleted uranium, the net consumption of plutonium would be about 90 kg per TWh, i.e. almost 80 % of the maximum value.

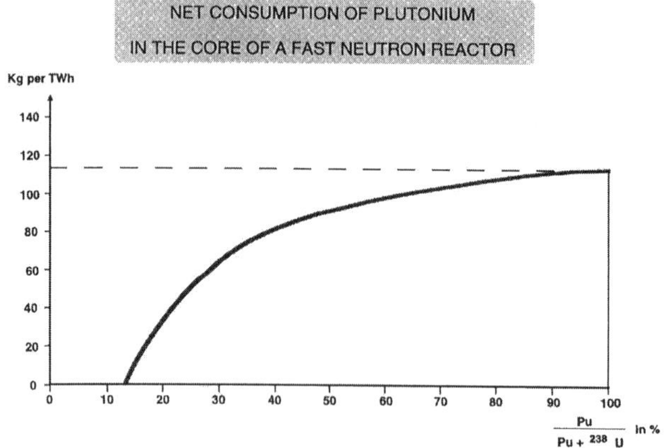

Fig. 7. Net Consumption of Plutonium in the Core of a Fast Neutron Reactor.

Just to give an order of magnitude - not of course to suggest that this could be done tomorrow - in a nuclear power system made up as three quarters of standard PWRs burning slightly enriched uranium fuel, and one quarter of fast neutron reactors designed for a maximum consumption of plutonium, there would be under normal running conditions no net production of plutonium.

Following the decision of the French government in 1992 to use Superphénix as an experimental tool to explore, under full scale conditions, the possibility to burn plutonium and other actinides in a fast neutron power plant, a research and development (R&D) programme extending over many years was prepared. Fuel assemblies containing neptunium, or with a higher plutonium to uranium ratio than the standard ones, were delivered to the reactor site as early as autumn 1996. Unfortunately all these preparations were to no avail, as the political decision was taken later on not to let the plant operate anymore.

3. Accelerator-Driven Systems[255]

The possibility to make strong neutron sources by means of spallation nuclear reactions, caused by high energy protons impinging on heavy nuclei, is not a new idea. It dates back half a century at least. Early 1950, the US Atomic Energy Commission approved a proposal from Ernest Lawrence to construct a linear accelerator with proton currents on the order of 50 mA at an energy of 25 MeV. The Mark 1, as the accelerator

---

[255] Below, I attempt to discuss accelerator-driven subcritical reactors, since I was asked to do so. Since accelerator-driven reactors were not a subject of active research and development during my professional years at the French *Commissariat à l'Energie Atomique* (CEA), I do not pretend to possess any particular familiarity with these systems.

was called, would generate a flood of neutrons and produce several isotopes of interest to the Commission, like plutonium or tritium. In parallel, design was proceeding on a huge accelerator called Mark II, to be housed in a tank 60 feet in diameter and 350 feet long, with an energy of 350 MeV and a current of 500 mA, which would require about 150 MW of electricity and could produce half a metric ton of plutonium a year. In 1952 the construction of the Mark I was completed at Livermore, a new site in California soon due to become a major thermonuclear research centre. Yet later that same year, the AEC concluded that production accelerators could not compete with production reactors. All plans for Mark II were deferred and Mark I was left to die a natural death.

At about the same time Bennett Lewis in Canada proposed to build an accelerator as a neutron source to produce ^{233}U for closing the thorium nuclear fuel cycle in CANDU heavy-water reactors. One accelerator with a 300 mA current of 1 GeV protons would suffice to feed twelve such reactors. Yet economic and financial considerations put a stop to this undertaking.

During the following decades up to now, studies of production accelerators for military or civilian purposes continued in various places, taking advantage of the great strides made in accelerator technology, but without any project being carried into effect, as far as I know.

About twenty years ago were proposed for the first time so-called hybrid systems, making use of a high energy (1 GeV) and high current (10 mA) proton accelerator to drive a subcritical reactor for transmuting minor actinides. Thereafter a large number of various schemes were investigated for the same purpose, in particular at the laboratories of Los Alamos and Brookhaven in the United States, at the Japan Atomic Energy Research Institute (JAERI) in Japan, in Russia, and elsewhere as well.

A further step was taken in the early 1990s, when Charles Bowman at the US Los Alamos laboratory, and later on Carlo Rubbia at the European Laboratory of Particle Physics (CERN, Geneva), put forward with great enthusiasm the concept of an "energy amplifier", whereby a hybrid system is specifically designed for the generation of electricity. Typically the latest design of Carlo Rubbia combines a proton accelerator (1 GeV, 30 mA or so) to a fast neutron subcritical reactor of 1500 MW thermal power, cooled by convection driven molten lead. At present such projects are very much in fashion, and some people view them as a way to offer a second chance to nuclear power at a time when it is faced with strong opposition in various countries.

As for hybrid systems[256], the extremely sophisticated accelerators, as well as the people who used them to make spectacular headway in our knowledge of elementary particles and their interactions, should be highly respected. Given the accelerator scientists' eagerness and ability, there seems undoubtedly to be scope to develop and improve further the technique of these accelerators towards achieving proton beams of larger and larger intensity. High energy and high current accelerators may readily find useful applications in many scientific and technical areas.

However, I would recommend extreme caution to those trying to find an opening for such machines by connecting them with subcritical reactors in hybrid systems. This represents a very challenging undertaking, which could lead to great disappointments to

[256] Here, I underline that I do not have any particular competence in the field of accelerators.

the extent that their expertise in the design, construction, operation and maintenance of nuclear reactors might be limited.

In any such project one of the key difficulties relates to the so-called window which forms the boundary between the accelerator and the reactor. It must comply with very severe and conflicting requirements. Its working lifetime is likely to be quite short in a system designed for power generation. Easy and efficient methods for checking up and controlling at all times the operation of the window and for replacing it before it fails are thus of prime importance. There is still a long way to go before well qualified solutions are at hand.

A drawback common to all high energy spallation sources of very fast neutrons is their strong anisotropy relative to the proton beam. The spatial distribution of the cascade of neutrons emerging from the target presents a pronounced forward peak, and the same holds true for the fissions they bring about, which is unfavourable with respect to the efficiency of the heat transfer system.

The technical characteristics of some current projects are not addressed here, since, in my opinion, some are wholly unrealistic. Many features of the preliminary conceptual designs existing today will very likely not hold out after further studies are carried out.

Nevertheless, a few general comments can be made. First of all, most present designs of hybrid systems call for subcritical reactors operating in a fast neutron spectrum, which is not surprising on account of the superior neutron economy achievable in that case. Thus the main issue is to discuss the pros and cons of fast neutron hybrid systems in comparison with fast neutron critical reactors.

The composition and the design of the medium where neutrons multiply by way of chains of fissions, and in particular the detailed characteristics of the nuclear fuel, are not a discriminating factor, as they can be chosen to be very similar, if not identical, in both cases. To be sure, the β value (fraction of delayed neutrons) of some higher actinides is low, so that they might not be suitable to constitute alone the fuel of a critical reactor. I do not see any reason, however, why those could not be associated in that case with other heavy nuclei possessing a high enough β value.

Sometimes one hears proponents of hybrid systems insisting on their running on thorium as their basic fuel, while claiming that they can thus avoid dealing with such a devil incarnate as plutonium. Such statements might be true, but they are only part of the picture. Thorium is not a fissile substance by itself, and to sustain a chain reaction of fissions in a medium mostly made of thorium, the presence of a fissile component is essential, be it ^{235}U or ^{233}U if plutonium is outlawed. Neutron capture in thorium gives rise to ^{233}U, so that a hybrid system can operate on the ^{233}U/thorium fuel cycle, just as a critical reactor can do. There does not seem to be any fundamental difference between resorting to ^{233}U and using plutonium.

One of the main arguments in support of the hybrid system is a safety one, namely that it would make possible to escape the Bethe-Tait curse of a power excursion. This is of course a very important issue. I fully agree that a decisive advantage would be granted to the subcritical reactor if it could be shown that any kind of accidental power surge is likely to occur with devastating consequences in a critical fast neutron reactor. But this is not the case.

A considerable amount of studies, tests and analyses, carried out in many countries, have led to the clear conclusion that there are ways to cope with this problem. To mention only our domestic situation, let me remind you that the French safety authorities have repeatedly declared that our fast neutron reactors are not inferior, with respect to their degree of safety, to the other French reactors, say the PWRs, though the technical questions are not all the same and must be dealt with separately in the different cases. In particular, they stressed - as did the French government itself - that the decision to shut down Superphénix had nothing to do with safety.

If people are not willing to agree to the safety of a fast neutron reactor, I doubt that they could be convinced that a slightly subcritical one is safe enough. A hybrid system will have to be operated with a value of k_{eff} close to 1. The margin with respect to criticality, which will anyhow evolve in the course of operation, is likely to be small enough for the safety authorities not to rule out the possibility that the reactor becomes critical some day and consequently to require appropriate measures to cope with this situation.

Apart from criticality accidents, the main safety issues, which must be dealt with for a fast neutron reactor, will also have to be treated on a subcritical one. For instance, the problem of the decay heat removal after shutdown will be the same in both cases, for a given power released in the nuclear fuel.

There is no difference either concerning the production of nuclear wastes. The quantity of fission products, which accumulate in the core, depends only on the total thermal energy released by the fissions in the fuel. Whether the reactor is critical or subcritical does not play any role.

Assuming that it is possible to solve the problems (including those related to safety) raised by the conjunction, within one single entity, of techniques as different as those of big accelerators on the one hand and those of nuclear reactors on the other hand, it is clear that the additional constraints involved in the case of the hybrid system will make it a more complex, more difficult, and consequently more expensive way to achieve what a standard critical reactor can do. This will be true whichever the purpose might be, whether it is electricity generation, plutonium consumption, actinide transmutation, nuclear waste burning, or any combination of them.

I cannot believe that such a device would ever lead to the generation of significant quantities of electricity under industrially viable and economically competitive conditions. Of course, reducing the scope of the undertaking to the sole purpose of destroying minor actinides and transmuting a few long-life fission products could only help by limiting the performances and the power of the required installation. But one can really question the interest in developing a new system for such a limited objective, whose necessity is by no means demonstrated.

Burning minor actinides is at any rate a minor problem, in comparison with that of burning plutonium itself, which accumulates at present at a tenfold larger rate. For sure it will be wise some day to take advantage of this valuable plutonium mine and to adopt measures to burn it. But the energy contained in the growing plutonium stockpile produced by existing reactors is already so large that its consumption can only take place in large nuclear power plants, which in my opinion have to be equipped with well-proven critical reactors.

A last word about the technique of accelerator driven neutron sources, to draw your attention to the fact that it is well fitted to the production of materials of interest for nuclear weapons. For example, a high energy and high current cyclotron is a rather compact machine, of relatively modest size. It would be the ideal tool to manufacture in an unobtrusive way substances of high military quality: plutonium from depleted uranium, ^{233}U from thorium or tritium from some light isotopes. As no fissile material, or very little of it, would be present to start with in the medium in which the accelerator driven neutron source is located, the quantity of fission products formed during the operation of the accelerator would remain limited. So would be the problems, which are raised by the intense radioactivity of the fuel irradiated in a production reactor.

References

Rémy Carle, *L'électricité nucléaire*, Que sais-je? Series n° 2777, PUF, Paris, 1993.

Jean Teillac, *Les déchets nucléaires*, Que sais-je? Series n° 2385, PUF, Paris, 1988.

Daniel Blanc, *La sûreté de l'énergie électronucléaire*, Que sais-je? Series n° 2032, PUF, Paris, 1991.

Maurice Tubiana and Michel Bertin, *La radiologie et la radioprotection*, Que sais-je? Series n° 2439, PUF, Paris, 1989.

P. Boulinier, *Les surgénérateurs*, Collection CEA, Synthèses Series, Eyrolles, Paris, 1987.

Georges Vendryes, *Superphénix, pourquoi?*, Editions Nucléon, Paris, 1997.

Chapter 16 Prospects for Accelerator-Driven Reactors: The Energy Amplifier

by
Bob van der Zwaan*

Introduction

The obstacles nuclear energy is currently facing are considerable and numerous. Today, it is difficult to see how its handicaps can be overcome. The future of nuclear energy is therefore uncertain. Nuclear energy might, however, become vital for mankind in the long run, for example as a means to mitigate problems such as global warming. Also, with respect to the availability of the planet's natural resources, nuclear energy possesses beneficial properties. A continued use of nuclear energy well into the 21st century would exhaust uranium reserves. Eventually, the use of fast neutron reactors would become imperative, since they are able to multiply the energetic capacity of natural resources by several orders of magnitude.

In 1993, Carlo Rubbia re-launched an old idea to develop a special kind of fast neutron reactor: an accelerator-driven, or hybrid, reactor. He baptised his design for such a reactor "energy amplifier". Rubbia's project would combine a particle accelerator and a "sub-critical" nuclear fission reactor into one integrated system (see figure 1). This "cross-fertilisation" between two different technologies would not only generate electricity, but could transmute long-lived radioactive fission products into shorter lived ones, as well as incinerate plutonium and long-lived minor actinides. The energy amplifier's fast neutron environment and its use of thorium as fuel would not only enable it to burn nuclear waste from current conventional reactors, but would itself generate considerably lower quantities of waste than those currently produced by nuclear power plants.

If these characteristics could be realised, an energy amplifier would address several of the main problems that nuclear energy is facing today. A hybrid reactor could perhaps respond satisfactorily to profound scepticism expressed against nuclear energy by the public. Nuclear energy, whose growth in global energy share in the 1970s and 1980s has recently started to stagnate, could possibly be given a new *élan* by developing hybrid reactors. Building an accelerator-driven reactor, however, is far from straightforward and

---

* Bob van der Zwaan is a physicist (CERN, Geneva) and an economist (Cambridge, UK). He has been researching on various aspects of nuclear energy at the *Institut Français des Relations Internationales* (IFRI, Paris).

many technical problems need to be overcome. What is the feasibility to develop such a reactor? This chapter assesses the merits of the energy amplifier and analyses some of the obstacles which need to tackled in order to allow its development. It argues that research into the many aspects of the energy amplifier should be intensified.

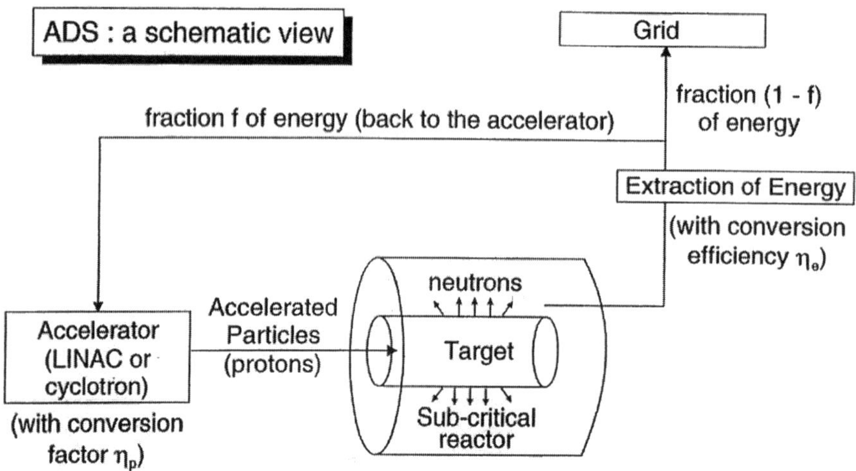

Fig. 1. A schematic view of an accelerator-driven system (ADS).
Source: M. Salvatores, CEA, 1998.

1. Reactor Safety

Although Rubbia's project has shifted from energy production to waste transmutation, its main ambition remains the development of a new generation of nuclear power reactors. Part of the electricity produced by his hybrid reactor (indicated by the fraction *f* in figure 1) is assumed to be fed back into the system, in order to operate the accelerator (see also figure 2). The fraction of the electric power used to run the accelerator would typically be of the order of 5 to 10 %. Since the remaining part (1 - *f*) constitutes electricity for consumption, Rubbia's accelerator-driven reactor is called an "energy amplifier".[257]

[257] See, for example, C. Rubbia and J.A. Rubio, *A Tentative Programme Towards a Full Scale Energy Amplifier*, CERN, CERN/LHC/96-11 (EET) and C. Rubbia, *The Energy Amplifier: a Description for Non-specialists*, CERN, CERN/ET/Internal Note 96-01. In other words, the name "energy amplifier" is derived from the gain factor that exists between the beam power and the total power produced, resulting from the

In a classical nuclear reactor, neutrons, freed by fissioning nuclei, induce other nuclei to break up. Conventional nuclear reactors are "critical", implying that fission processes in the reactor fuel can produce just enough neutrons to sustain a continuous chain of fission reactions. One of the outstanding differences with a conventional nuclear reactor is that the energy amplifier is sub-critical. It therefore cannot sustain a chain reaction without the help of the accelerator. In such a hybrid reactor, fission reactions do not create enough neutrons to provoke a continued fission sequence. The accelerator provides the "missing" neutrons for sustaining a fission chain reaction. The sub-criticality of the energy amplifier makes it what is called by some scientists "inherently safe". If at any moment a technical problem would occur, or if the reactor would get overheated, the accelerator can simply be switched off. In fact, in Rubbia's design it would be switched off automatically. This results in an abrupt stop of the chain reaction. Accidents like that of Chernobyl can therefore not occur.[258]

The accelerator of an energy amplifier would need to boost protons up to an energy of about 1 giga-electronvolt (1 GeV) with a central current (intensity) of about 30 milli-amperes (30 mA), in a possible range from 10 to 200 mA. This high-energy, high-intensity, proton beam would be targeted into the core of the reactor on the liquid lead functioning as coolant. The collisions between the high energy protons and the lead nuclei would create, by spallation, an avalanche of neutrons. These are the neutrons required to "bridge the gap" and provoke a fission chain reaction.

In order to control the chain reaction in a conventional reactor, specially designed rods can be lowered down into the reactor core, capturing neutrons. Thereby, one controls the evolution of the chain reaction and avoids an uncontrolled expansion of it. One of the causes of the accident in the Chernobyl power plant was an improper design of the control-rod mechanism. This contributed to the reactor's malfunctioning and provoked an over-amplified chain reaction, which led to the catastrophic melt-down of the reactor core.[259] The energy amplifier does not need control rods.

The molten lead has an excellent heat absorption capacity and is therefore a very suitable coolant. Rubbia's design states that, unlike in a light or heavy water reactor, where water serves as coolant and pumps are needed to carry the heated water outside the reactor vessel, no pumps are needed in the energy amplifier vessel. The natural convection properties of the liquid lead are used to transfer the heat from within the core to the coolant surface in the reactor vessel. Heat exchangers vaporise a secondary coolant, such as water, subsequently used to feed turbines for the production of electricity.

---

externally driven nuclear fission cascade. In terms of the law of conservation of energy, the name "energy amplifier" is perhaps somewhat confusing.

[258] Some specialists state, however, that the energy amplifier might encounter problems in the residual heat removal, once the system has been switched off.

[259] For a more extensive description of the Chernobyl accident, as well as a complementary concise description of the energy amplifier, see, for example, B. van der Zwaan, *L'Energie nucléaire au 21e siècle : Enjeux de sécurité*, IFRI, Paris, 1999.

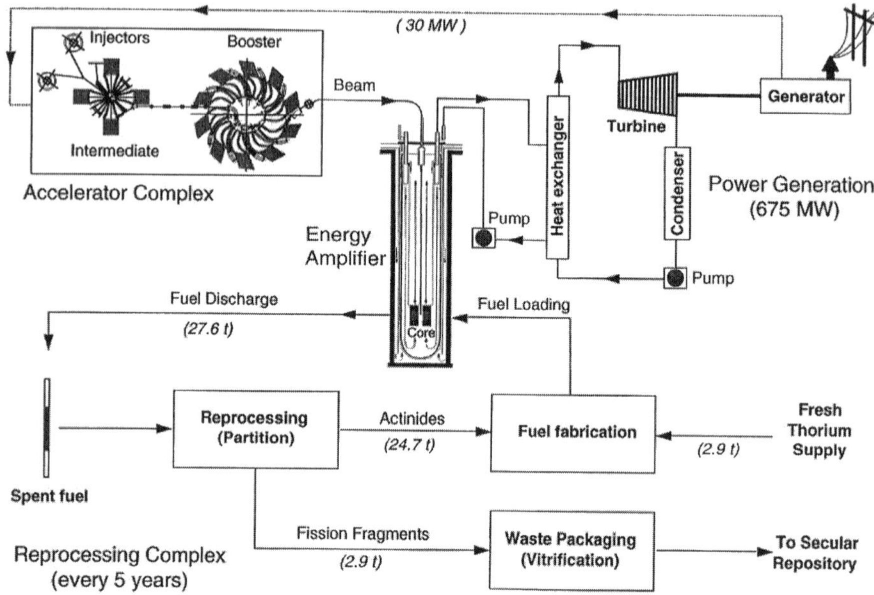

Fig. 2. A schematic view of the general lay-out of the energy amplifier complex. *Source: C. Rubbia, CERN, Geneva, 1995.*

2. Accelerator Technology

The interesting aspect of the Rubbia design is that the required accelerator technology seems, at least in principle, to be available. This is mainly a result of the enormous progress that has been achieved in this area of physics since the end of World War II. Many different types of accelerators have been developed at numerous research institutes around the world, notably at the European Laboratory for Particle Physics (CERN) in Geneva. Two types of accelerators can be used for the hybrid reactor: a linear accelerator (LINAC) or a circular accelerator (e.g. a cyclotron).

Both accelerators are already able to reach the required energy. With respect to the intensity of the beam some technological progress still needs to be made. Circular accelerators, having the advantage above the linear ones that they are relatively small and thus easier to install, can reach intensities of just about the required level of 30 mA. They should be subjected to further research and development in order to more adequately confine the beam and thereby increase its intensity. Linear accelerators can already reach intensities of the order of 100 to 200 mA, but can basically only do so in a pulsed mode. A significant technological constraint has still to be surmounted to achieve an approximately continuous operation of the proton beam.

A system using a proton beam of 15 mA with an energy of about 800 mega-electronvolt (800 MeV) could drive a unit of approximately 200 mega-watt of electrical power (200 MWe). A circular accelerator would do very well for this purpose. A high-current linear accelerator might be used for establishing an energy production system with 1200 mega-watt electrical power (1200 MWe), comparable to the capacity of current pressurised water reactors.

The production of energy is Rubbia's long-term goal, but the energy amplifier's capability to transmute long-lived radioactive isotopes into radioactive elements whose half-life is substantially shorter is his project's current first phase focus. Several projects, in particular in the US, have in the past been investigating the possibilities of transmuting waste by using a particle accelerator. Ideas involving the use of accelerator technology for various purposes other than fundamental research have been addressed as early as the 1950's.[260] Japanese scientists have been doing research on these ideas and the Soviets (Russians) have been showing interest in related research for decades.

Although projects attempting to develop the principle of incorporating an accelerator in a power generation system have regularly emerged in the past, most of them seem to have been initiated with the main purpose of transmuting long-lived radioactive waste. At the US Los Alamos Laboratory, Charles Bowman has designed a hybrid reactor with a main ambition to incinerate plutonium, minor actinides and fission products. The Rubbia project at CERN has proposed to combine transmuting waste with generating electricity, in a way which previously has never been put forward so ambitiously.

3. Thorium Fuel and Plutonium Reduction

For various reasons, thorium is planned to become the fuel for the energy amplifier. The commonly used conventional "uranium fuel cycle" uses uranium 235 as natural primary fissile material and uranium 238 as natural primary fertile material, while producing plutonium 239 as artificial secondary fissile material. Plutonium 239 can be recycled from the spent fuel, and subsequently used as primary fissile material. The proposed "thorium fuel cycle" uses thorium 232 as natural primary fertile material, while producing uranium 233 as artificial secondary fissile material. Uranium 233 is required to be recycled from the spent fuel, in order to constitute the primary fissile material.[261]

One of the main advantages of using thorium as fuel is that in a thorium-based reactor, compared to a conventional uranium reactor, the production of plutonium (which in a uranium reactor results from uranium 238 nuclei capturing neutrons) is considerably reduced. The reason is that thorium is a lighter element than uranium, i.e. the atomic number Z, or number of protons in the nucleus, is lower for thorium ($Z = 90$) than for

[260] See, for example, the preceding chapter in this book, by Georges Vendryes.

[261] Note that, to start the first cycle, one needs to use uranium 235 or plutonium 239 as primary fissile material, since uranium 233 is initially not available. It is only after the first cycle that uranium 233 can be used as fissile material in subsequent cycles.

uranium (Z = 92).[262] Plutonium is a highly radio-toxic material and constitutes a large threat for non-proliferation regimes.

Rubbia wants to combine a (thorium fuelled) fast neutron reactor with an accelerator, by which the amount of plutonium produced could be even further reduced. In fact, in Rubbia's design plutonium can be used as part of the fuel. The accelerator produces fast neutrons which are able to incinerate plutonium. Thereby, the amount of plutonium going into the reactor could possibly be larger than the amount going out, implying a net reduction of plutonium.[263] Rubbia claims that, with a set of energy amplifiers, around the middle of the next century the entire stocks of existing civil and military plutonium could in principle be eliminated.[264] The energy amplifier would also be able to promote the destruction of other dangerous fissile materials.

In the energy amplifier, plutonium is transmuted into lighter elements. This so-called "accelerator-based conversion" of plutonium could, in theory, become an alternative to the "dual track" disposition strategy for excess stocks of (both civil and military) plutonium, such as proposed by the Committee on International Security and Arms Control (CISAC) of the American National Academy of Sciences (NAS). This plutonium disposition strategy consists of either mixing plutonium with waste from nuclear power plants and vitrifying the mixture, on the one hand, or producing and burning Mixed plutonium uranium Oxide (MOX) fuel, on the other hand.[265] The former method immobilises plutonium by combining it with long-lived and highly radioactive waste and fashioning it into glass or ceramic logs. Incinerating plutonium via MOX fuel is the alternative, in which plutonium is embedded in uranium and subsequently used as fuel in conventional reactors.

In both cases, the plutonium is embedded in highly radioactive materials and is therefore difficult to recover and re-use in weapons; the plutonium attains the so-called "spent fuel standard". Both cases would require a final solution such as underground storage, in the US for example investigated at Yucca mountain. The American NAS described the incineration of plutonium via accelerators as something both technically

---

[262] Thus, for thorium more neutron captures - each such a capture leading by a subsequent b particle emission to an element with a one integer higher Z - are required to reach plutonium (Z = 94). Thereby, thorium "reaching" plutonium is less likely than uranium reaching plutonium.

[263] C. Rubbia, S. Buono, E. Gonzalez, Y. Kadi and J.A. Rubio, *A Realistic Plutonium Elimination Scheme with Fast Energy Amplifiers and Thorium-Plutonium Fuel*, CERN, CERN/AT/95-53 (ET).

[264] Recall that the inventory of civil plutonium is estimated to be about 900 tonnes in addition to about 250 tonnes of military plutonium. See: D. Albright, F. Berkhout and W. Walker, *Plutonium and Highly Enriched Uranium 1996*, SIPRI, Oxford University Press, 1997. The total amount of plutonium increases steadily, since power reactors are, world-wide, discharging about 70 tonnes of plutonium each year in the form of spent fuel.

[265] *Management and Disposition of Excess Weapons Plutonium*, CISAC, NAS, Washington, 1994. See also: J.P. Holdren, "Work with Russia", *The Bulletin of the Atomic Scientists*, March/April 1997. Which of the two methods has to be given preference to has for various years been a subject of discussion. For the moment, both options are being kept open. See, for example, E.S. Lyman and P. Leventhal, "Bury the Stuff", *The Bulletin of the Atomic Scientists*, March/April 1997, and J.P. Holdren, "Dangerous Surplus"; L.J. Carter, "Let's use it"; A. Makhijani, "Let's Not", *The Bulletin of the Atomic Scientists*, May/June 1994.

and financially not foreseeable in the near future.[266] Indeed, it might take decades before an energy amplifier prototype can be developed. Building a whole set of energy amplifiers, required to eliminate a considerable amount of plutonium, would take much longer. The actual plutonium stocks are too big for the energy amplifier to play, in the near future, a significant role in their destruction. Therefore, the current plutonium disposition efforts should in no way be put into question.

Since the energy amplifier uses thorium, it could open an entirely new approach to generating electricity. The abundance of thorium has been a matter of some controversy: critics claimed that thorium resources in the earth crust are not so much larger than those of uranium.[267] Numerous experts, however, agree that thorium is about 3 times more abundant on earth than uranium.[268] Continuing with current levels of consumption, the currently known uranium resources will run down in about 100 years (see for an account of the relationship between uranium supply and market price, and the potential for the recovery of uranium from seawater, the chapter on reprocessing by Garwin). In addition, alternatives for the use of fossil fuels need to be found in order to curb the green-house effect. Both factors would make the energy amplifier, based on a thorium fuel cycle, a particularly advantageous tool.

Mining thorium ore seems radiologically to be more hazardous than uranium ore, provoking critique from environmentalists.[269] The thorium cycle imposes higher potential radiological risks for the personnel working with the various materials involved, than the uranium cycle. One would possibly need to rely on remote operation for both ore processing as well as fuel fabrication and/or recycling. On the other hand, thorium ore extraction presents no hazardous radon effect, and, in the long run, the residues of thorium mining have a lower radiotoxicity than those from uranium mining.[270] If thorium would be used as a future energy resource, new methods of mining, milling, fuel fabrication and reprocessing would need to be developed. The fact that new techniques are required, makes cost estimations particularly difficult. There exists a considerable uncertainty about the level of extraction costs: the investments required could possibly prove substantial. A beneficial aspect is that, supposedly, one needs to mine several orders of magnitude less thorium than uranium ore. This would produce considerably less amounts of mining and milling wastes.

Another attractive feature of using thorium instead of uranium, is that no difficult, time consuming and expensive enrichment and conversion procedures are required. Whereas natural uranium consists of two distinct isotopes (uranium 235 and

---

[266] See, for example, *Management and Disposition of Excess Weapons Plutonium, Reactor Related Options*, CISAC, NAS, Washington, 1995. See also: M. Moore, "Plutonium: The Disposal Decision", *The Bulletin of the Atomic Scientists*, March/April 1997.

[267] M. Pavageau, M. Schneider, *The Rubbia TABS, Solutions or Illusions?*, WISE-Paris, January 1997, p. 16.

[268] G. Charpak and R.L. Garwin, *Feux Follets et Champignons Nucléaires*, Editions Odile Jacob, 1997.

[269] M. Pavageau, M. Schneider, *op.cit.*, p. 12.

[270] Using thorium instead of uranium also has advantages with respect to the long-lived waste produced in a nuclear reactor. The immediate radiological impact on personnel working with thorium fuel, however, might be considerably larger. See: J.P.Schapira, "Le rôle potentiel du cycle du combustible à base de thorium" in: R. Turlay (ed.), *Les déchets nucléaires*, Les Editions de physique, 1997.

uranium 238), of which only uranium 235 is fissile, natural thorium consists only of one isotope (thorium 232), which is fertile. Uranium used in conventional reactors normally needs to be enriched from the natural 0.7 % to about 3-4 % of uranium 235. Thorium just needs to be mixed, initially, with some fissile material like plutonium in order to constitute a proper fuel in an energy amplifier. At a later stage, the production of fissile uranium 233 would make the energy amplifier operable without requiring an additional fissile material.

4. Thorium Fuel and Non-Proliferation

Much research has been, and is being undertaken with respect to the feasibility of using the thorium fuel cycle. In a design by Alvin Radkowsky, based on a thorium version of the Soviet VVER reactor ("VVER" being the Russian abbreviation for "pressurised normal water reactor"), one could reduce the amount of produced plutonium by about 80 %.[271] Moreover, the isotopic composition of the plutonium coming out of a thorium cycle is such that its use in a nuclear bomb renders the explosive yield of the bomb less powerful. In addition, the presence of a relatively large amount of plutonium 238 in the plutonium composition causes a large heat emission so that it is more difficult to handle and therefore less accessible for use in nuclear weaponry.

The absence of a primary fissile material, *a priori*, in the thorium fuel cycle is a major obstacle for creating a cycle *solely* based on thorium. Various solutions proposed in the past used enriched uranium or plutonium as an additional primary component. The use of these materials, which need to be enriched or separated first, make a thorium reactor still sensibly proliferative. Radkowsky claims that thorium based light water reactors can now be constructed, which would disassociate unambiguously civilian from military nuclear power.

As was shown, in the energy amplifier the produced uranium 233 needs to be recycled. One of the drawbacks of this is that uranium 233 is one of the most fissile isotopes known and hence constitutes, in principle, a hazard for non-proliferation policies. In other words, although having, as a non-proliferation advantage, the production of less plutonium and possibly even the incineration of stocks of excess plutonium, the thorium cycle brings about a proliferation peril by the production of uranium 233, since this isotope plays a role comparable to that of plutonium 239 in the uranium cycle and has strong explosive potentials. It is readily separable from thorium and can, in principle, be used in nuclear weapons. Rubbia's project is therefore said to solve one proliferation threat while creating another. Some scientists state, however, that the thorium cycle is still much more proliferation resistant than the uranium cycle.[272] One

---

[271] The concept of Radkowsky is a so-called "VVERT" reactor, where the T refers to the fact that it is a VVER reactor using *thorium* as fuel. See A. Galperin, P. Reichert, A. Radkowsky, "Thorium Fuel for Light Water Reactors - Reducing Proliferation Potential of Nuclear Power Fuel Cycle", *Science and Global Security*, vol.6, no.3, 1997, pp. 265-290.

[272] A. Galperin, P. Reichert, A. Radkowsky, *op. cit.*

of the reasons would be that uranium 233 is produced along with uranium 232 and is difficult to separate from this isotope. Because of the radioactivity of uranium 232, uranium 233/232 is difficult and dangerous to handle. The fabrication of bombs would therefore be tedious.[273]

Another proliferation drawback of the energy amplifier is the use of a particle accelerator. With an accelerator, very suitable for the production of neutrons, one can produce a wide variety of different elements. A rogue state possessing such a machine could thus produce radioactive isotopes and fabricate radiological weapons, or produce fissile elements for the construction of atom bombs. These risks have so far been receiving little attention, but it is important to underline that accelerators have certain disadvantages in terms of non-proliferation.[274]

5. Technical Obstacles

The physics principles of the energy amplifier concept all seem to be realistic. It is on the technical side, where many difficulties need to be solved. There are a number of obstacles, mainly on the engineering level, which need to be addressed in order to develop and build an energy amplifier. In the eyes of many sceptical reactor physicists, these technical obstacles are too numerous. They think there is ample reason to believe that the realisation of a Rubbia-type energy amplifier, if it will ever see the daylight at all, will take many years at least, if not decades. Of course, combining technologies as different and complex as that of accelerators and fission reactors necessarily implies enormous scientific challenges. Certainly, a lot of research is required before one can start constructing an industrial prototype of the energy amplifier.

The design of the energy amplifier specifies that the reactor vessel should be about 6 m in diameter and 30 m high (see figure 3). This large size imposes demanding requirements on the resistance of the vessel against the fluid lead cooling liquid. The construction should be able to bear a high pressure: designing and constructing such a vessel might not be straightforward. If one had to start from zero in developing the use of lead as a coolant, researchers would be occupied for at least a decade.[275]

---

[273] The radioactive decay of uranium 232 produces gamma rays of 2.6 MeV. The contamination of uranium 233 by uranium 232 makes the production of bombs thereby difficult for two reasons. First, the gamma radiation produced is such that in handling the material one would receive lethal doses of radiation in short time spans. Second, it is difficult to deal with the heat produced by the presence of this material in a bomb.

[274] This raises the difficult question whether particle accelerators should perhaps be put under (IAEA) safeguards.

[275] Carlo Rubbia recognises this as well. See "Carlo Rubbia; protons, neutrons, plomb et megawatts", *La Recherche*, 302, October 1997.

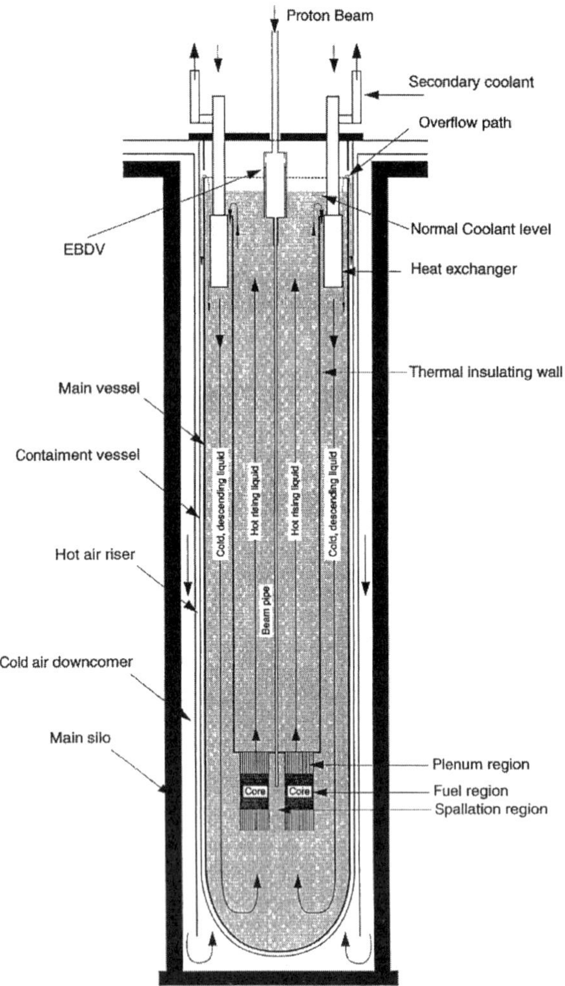

Fig. 3. A side view of the energy producing reactor vessel of the energy amplifier.
Source: C. Rubbia, CERN, Geneva, 1995.

That is why Rubbia would like to use the expertise available in Russia on lead coolants. The Russians have been using lead or lead-bismuth mixtures to operate reactors in submarines. The possibility of collaborating with the Russians, and perhaps buying their lead technology, is being explored. Several sources indicate, however, that the Russians have had difficulties with these reactors in the past. It is therefore uncertain whether this technology transfer is going to be fruitful. It is uncertain whether the use of natural convection would suffice as a means to transport the heat in the vessel. The corrosion of the vessel, and other reactor materials, by the liquid lead coolant has to be intensively investigated. This could pose all sorts of unexpected problems.

Another difficulty would be the recharging of fuel elements, which is not easy in a lead environment. Also, detailed studies need to be performed with respect to the radioactivity induced by fast neutrons in the various reactor materials.

One of the key technologies to be developed is the window between the accelerator and the reactor core. This window is designed to be made of tungsten. On the one hand, it needs to be sufficiently transparent to let the protons pass to get into the reactor vessel, while, on the other hand, it needs to be resistant enough to retain, at high temperatures, the pressure of the liquid lead in the vessel. The window, like the reactor vessel as a whole, needs to resist corrosion by the liquid lead. In addition, it needs to be strong enough to bear a constant high current and high intensity proton beam. A suitable design can probably be found, but it would undoubtedly need to be replaced regularly, e.g. once per year, since it would inevitably suffer damage from the proton beam. The difficulties with respect to the mechanism for replacing the window still need to be handled. The transition from the vacuum in the accelerator to the high pressure in the vessel is, in some respects, the Achilles' heel of the energy amplifier.

It is claimed by various experts from the nuclear energy industry that the costs to investigate and develop Rubbia's ideas might prove enormous. First of all, they claim, it is premature to give detailed cost estimates since too many vital elements are still missing in his proposal. The investment required for developing the whole system at once is estimated by some experts to run into tens of billions of ECUs (euros).[276] The capital needed to build a prototype seemed initially to be underestimated. One also needs to take into account that thorium is not yet used as fuel on a large scale, so that the industrial technologies currently used for mining, milling and fuel production have to be adapted. It is *a priori* not guaranteed that the energy amplifier will be able to sustain a competitive electricity price. Rubbia, backed by some economists, claims however that the energy amplifier electricity price could become very competitive indeed.

6. Waste Transmutation

Since the amount of research and development, in different fields of study, is too large to be undertaken at once, Rubbia's project focuses for the moment on using the accelerator for transmuting radioactive waste and incinerating plutonium. Rubbia has admitted that developing all aspects of the entire machine at once seems not sensible. Therefore, an incremental approach to the energy amplifier project is being taken, allowing solution of the various unsolved questions one by one. Figure 4 shows an example of the result of the transmutation principle - the project's current main orientation - on the reduction of the radiotoxicity of waste from conventional uranium based reactors.

---

[276] *Draft STC Opinion on Nuclear Energy Amplifier (with annexes)*, Euratom Scientific and Technical Committee, August 1997.

The Euratom Scientific and Technical Committee has advised the European Commission to finance two well-defined topics within the energy amplifier project as a whole. Thereby, it approves the actual incremental approach to the project. It supports continuation of research on the development of thorium matrix fuels for plutonium management and actinide minimisation, and the development of accelerator-driven fast neutron reactors for waste incineration.[277]

Transmutation of nuclear waste in a hybrid system, if successful, could alleviate a major problem that the electro-nuclear industry is dealing with, and an enormous threat with respect to global security. If Rubbia's machine should not succeed in producing electricity, it could just be used as a waste transmuter or plutonium incinerator. Such apparatus could be placed on nuclear power plants and solve the waste problem on-line.

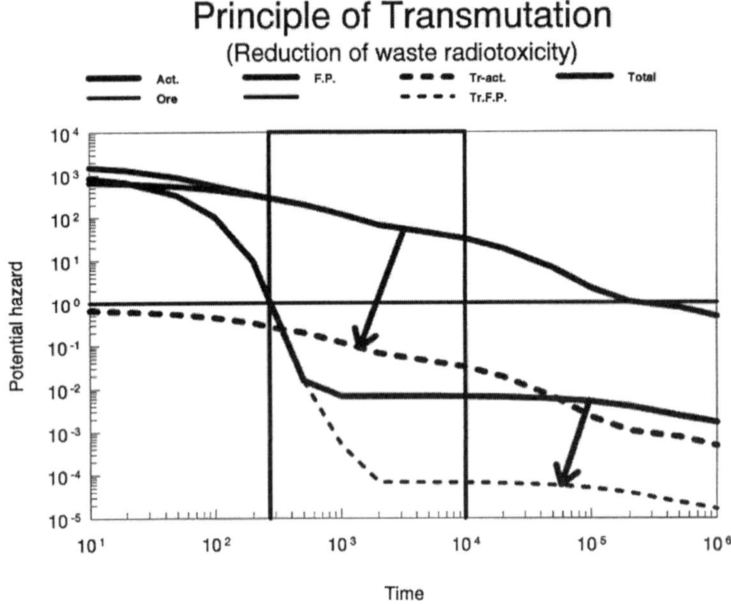

Fig. 4. An example of the transmutation principle. The upper arrow shows the result of actinide transmutation, the lower arrow indicates the result of fission product transmutation, both in terms of the reduction of the potential hazard.[278]
Source: H. Gruppelaar, ECN, The Netherlands, 1998.

The advantages of an energy amplifier using thorium with respect to nuclear waste are numerous. Not only does the use of thorium in such a hybrid reactor produce substantially less plutonium, but the same is true for minor actinides such as americium,

[277] *Draft STC Opinion on Nuclear Energy Amplifier, op. cit.*

[278] The potential hazard is expressed in arbitrary units, with respect to the horizontal line representing the hazard of natural uranium ore.

curium and neptunium. The production of minor actinides, with very long half-lives, is 2 to 3 orders of magnitude lower than that encountered in actual conventional reactors.[279] Similarly to the case of plutonium, a hybrid system could contribute to incinerating minor actinides produced in current nuclear power plants. The main reason for this is that in a fast neutron environment, such as that of the energy amplifier, long-lived minor actinides become fissile and can thus be burnt. In other words, the energy amplifier would be able to transmute the wastes from the current generation of nuclear reactors, while it would produce itself only small amounts of long-lived radioactive waste.

At CERN, it has been shown that a hybrid system could also efficiently transmute various fission products, such as technetium 99 and iodine 129. Another advantageous property of the energy amplifier, which derives from the fact that thorium is used as fuel, is that, during the first 10 000 years, the radiotoxicity of spent thorium fuel is 10 to 20 times lower (depending on the time evolved from the fuel discharge) than that of spent fuel from a uranium based reactor.[280] Because, in an iterative recycling process, plutonium becomes gradually "polluted" by non-fissile plutonium isotopes, one cannot continue recycling *ad infinitum*. For the thorium cycle, the reprocessing of uranium 233 does not impose similar constraints, so that one can, in principle, recycle indefinitely. In practice, this means that much higher levels of the amount of fuel can actually be used (burnt) in the case of the thorium cycle. This constitutes an advantage in the light of the amount of radioactive waste produced.

To underline the importance of the energy amplifier project with respect to waste transmutation, one has to realise that the actual solutions for intermediate surface storage of nuclear waste, and future underground storage, are very expensive. The same is true for storing and safeguarding both excess weapon grade plutonium and spent fuel standard plutonium. Addressing nuclear waste management and plutonium disposal in alternative ways is therefore worthwhile. A French law of 1991 promotes transmutation as one of three possible axes along which to proceed for dealing with nuclear waste. A French parliamentary commission has recently advised the French minister of research to support the project of Rubbia, if possible in a European programme gathering all parties concerned.[281] Although no firm decisions on extensive research programmes have been taken so far, it is possible that, in the near future, European efforts in the field of hybrid reactors will be enhanced. The International Atomic Energy Agency (IAEA) organised a conference, in the Fall of 1997, dedicated to accelerator based systems, thereby demonstrating its interest in this matter.[282]

Developing an accelerator-driven waste transmuter is unlikely to resolve the nuclear waste problem entirely and it would probably not render deep underground

---

[279] "Les systèmes hybrides de production d'énergie", H. Nifenecker in *Les déchets nucléaires*, R. Turlay (ed.), Les éditions de physique, 1997.

[280] "Le rôle potentiel du cycle du combustible à base de thorium", J.P. Schapira in *Les déchets nucléaires*, R. Turlay (ed.), *op. cit.*

[281] C. Birraux, *Contrôle de la sûreté et de la sécurité des installations nucléaires*, Tome I: Conlusions du rapporteur, No. 3491, Assemblée Nationale, April 1997. See also: "France urged to head Rubbiatron efforts", *Nature*, Vol. 386, p. 426, 3 April 1997.

[282] V. Arkhipov, "Future Nuclear Energy Systems: Generating Electricity, Burning Wastes", *IAEA Bulletin*, 39/2/1997.

storage unnecessary. If, however, such a transmuter could be developed, within the coming decade or so, and if it should prove to be an efficient actinide burning machine, capable of destroying long-lived radioactive isotopes, it could mitigate the nuclear waste problem substantially and make (reversible or irreversible) deep underground storage less urgent and expensive.

Deep storage seems to be fairly safe for at least several thousands of years. Beyond that time, scientists still question whether the buried material could possibly diffuse to the surface. Another problem is that once human memory will have faded and future generations will have forgotten about their ancestor's deep underground storage, they could possibly start digging at depository sites and accidentally encounter nuclear waste. If with a Rubbia-type machine the effective lifetime of radioactive waste could be substantially reduced, one would possibly need to rely less on deep underground storage. The safety time limits of such storage could perhaps become sufficient. Since it will take still many years before the final deep storage decision will be taken, and at least 10 years - but likely much longer - before the first geologic depository will be ready, this delay can appropriately be used for continuing research into the development of accelerator based waste incineration.

Conclusion

Undertaking research and development in accelerator-driven waste transmutation, such as with an energy amplifier, is justified, since it constitutes an attempt to solve a major problem national governments have to deal with sooner or later. Whether governments decide to abandon nuclear energy or not, the nuclear waste problem has to be tackled. By developing accelerator-driven reactors for waste incineration, one keeps open, at the same time, the possibility for establishing a new "accelerator-driven option" to produce electricity. Given the dangerous increase of the greenhouse effect, due to excessive anthropogenic carbon-dioxide emissions, and given the fact that an energy amplifier - and nuclear energy in general - does not emit carbon dioxide, research in such radically innovative reactors should be encouraged.

In order to achieve that, in 2050, more than 50 % of total commercial energy is supplied by carbon-free resources, it is important to undertake extensive research and development into *all* carbon-free energy opportunities. Such a strategy could avoid a radical, possibly catastrophic, climate change. Carbon-free energy resources such as decarbonised fossil fuels, photo-voltaics, biomass, hydropower and wind should be investigated much more intensively than is being done at present, and major efforts should be undertaken to establish higher levels of energy efficiency, in order to achieve a radical change in the use of energy.

In my opinion, however, one has to be wary of an overshoot-reaction vis-à-vis nuclear energy. It could prove unwise to abandon nuclear energy, since it is today, together with hydropower, the only carbon-free energy resource which produces significant amounts of energy. Of these two, only nuclear energy has the potential, in principle, to be expanded considerably. Research should be intensified in the nuclear field, such that solutions can be brought forth to its main obstacles: radioactive waste, reactor safety and nuclear proliferation. Research and development into the feasibility of an energy amplifier is one of the ways by which such solutions could be found, and should therefore be promoted.

Of course, it is far from sure that an energy amplifier could provide competitive energy prices in the near future. Given an inevitable exhaustion or near-depletion of fossil fuels within 50 to 100 years, and even of uranium stockpiles, in a similar time scale (unless breeder reactors can be employed massively, or uranium can be extracted from sea-water), however, such competitivity could be reached on the long run.

The properties of the energy amplifier with respect to reactor safety, radioactive waste transmutation and weapon materials elimination might restore the image of nuclear energy in public opinion. Given the capacity of nuclear energy to reduce carbon dioxide emissions, and thereby global warming, and given the energy resources which would become available by using reactors such as energy amplifiers, nuclear energy might even increasingly satisfy the criteria for a sustainable development. Thus, hybrid reactors could provide a revived interest in nuclear energy in the 21st century. Especially interesting with respect to sustainability would be a symbiosis between advanced conventional reactors, advanced breeder reactors and accelerator-driven transmutation reactors.[283] In fact, the role which evolutionary reactors (being closely related to current designs, but possessing higher safety qualities) could play, should not be underestimated.

Therefore, pushing further the research initiated by Rubbia is important, even if it is reasonable to assume that an accelerator-driven system could only become feasible in the long run. Indeed, the technical obstacles to be solved - some of which have been mentioned above - are numerous, especially if one has the high ambition to produce electricity. The obstacles should be dealt with in a step-by-step approach. The project's aim should first of all be limited to addressing mainly waste transmutation and plutonium incineration. If that proves fruitful, the project would already imply a major scientific break-through. Results on the transmutation of fission products have shown that substantial progress can be achieved in this direction. At a later stage, an energy amplifier capable of supplying electricity, essentially of the same design as a pure waste burner, could be developed. Technical obstacles, or uncertain economic payoff, should not prevent scientists from doing the relevant research. Scientific obstacles and challenges should be the reason for continued research in all fields, including that of hybrid reactors.

---

283 R. Chawla, "L'énergie nucléaire dans un contexte global", in *Face aux énergies fossiles, quelle place pour l'energie nucléaire?*, Actes de la Journée du CUEPE, Geneva, April 1998.

Chapter 17 The Risk of Proliferation and International Safeguards

by Georges Le Guelte*

Introduction

In the past, the view has been expressed that because nuclear deterrence has prevented a conflict during the Cold War between the most irreconcilable enemies, it would also prevent wars between any other adversaries. If all countries could avail themselves of nuclear arsenals, military confrontations would be excluded.[284] That opinion was expressed once more after the Indian and Pakistani tests in May and June 1998. If this theory would be right, the non-proliferation policy would be a big mistake. However, it supposes that nuclear weapons can only be used as the instruments of a deterrence strategy. History, since 1945, demonstrates the contrary.

After the beginning of the Cold War, in 1947, all experts were convinced that nuclear weapons would be utilised in the forthcoming conflict. Such a conflict seemed to be inevitable. They debated whether nuclear weapons should be used by air forces against cities, to demoralise the population, or by ground forces against large concentrations of troops.[285] In fact, their actual use has several times been considered seriously: in Korea or against China in the early fifties, before MacArthur was dismissed; in Indochina in 1954, after a request by the French government. In 1956, during the Suez crisis, the USSR threatened to use them against France or the UK, if they would not withdraw their forces from Egypt (although, at that time, the UK had already tested nuclear devices). In all cases, the decision not to make use of nuclear weapons was not based on their existence or their nature, but on political considerations.

The development, after 1953, of thermonuclear weapons - which are at least a thousand times more powerful than fission devices - and the deployment, after 1958, of long range missiles, generated the strategy of deterrence, not as a substitute for the strategy of use, but in addition to it. In fact, the only situation in which the nuclear arsenals were used as an instrument of deterrence to prevent a war was the Cuban missile crisis in 1962. An analysis of that episode demonstrates that deterrence is a very delicate strategy, requiring both technical and political conditions which are not easily met.

It must be emphasised that the mere availability of nuclear arsenals to both opponents may induce them carefully to avoid any kind of conflict. After 1962, the status of Berlin was no longer challenged by the USSR, and the Oder-Neisse border was never

---

* George Le Guelte is a former Secretary of the Board of Governors of the International Atomic Energy Agency (IAEA), and a former Deputy Director of International Relations in the French Atomic Energy Commission (CEA). He is the author of *Histoire de la menace nucléaire*, Hachette, Paris, 1997.

[284] See, for example, Kenneth N. Waltz, *The Spread of Nuclear Weapons: More may be better*, International Institute for Strategic Studies (IISS), Adelphi Series, n°171, Autumn 1981.

[285] See Lawrence Freedman, *The Evolution of Nuclear Strategy*, MacMillan, London, 1989.

put anymore into question. Wars only erupted between the East and the West in areas less important to the superpowers, and through allied forces (such as in Vietnam, Angola, Yemen, Afghanistan). US and Soviet soldiers were never directly engaged against each other in an open conflict. Such a careful attitude, however, does not derive automatically from the possession of nuclear weapons: the tests conducted by India and Pakistan did not prevent casualties on both sides in skirmishes on the border, and did not solve their dispute over the status of Kashmir.

If, in spite of all precautions, a crisis cannot be avoided, nuclear deterrence may prevent a war only if a number of technical and political conditions are met. Each side must be persuaded that a strike would be immediately and unavoidably followed by a deadly retaliation. If that retaliation is not certain, each of the opponents may feel that all the defence capabilities of the enemy can be eliminated by a surprise attack, and that a costless and definitive victory is possible.

Therefore, deterrence requires the availability of weapons which cannot be eliminated by a first strike (nuclear submarines, for example, since they are undetectable), and which will permit retaliation even if all other weapons have been destroyed. In addition, it requires a very sophisticated system of control and command, assuring that only the highest political authority can decide on the use of nuclear weapons, and that system must be able to survive a first nuclear strike. Deterrence demands also a wide detection network, preferably by satellites, to be informed immediately when a missile is launched by the opponent. It depends also to a large extent on distances between the enemies. If they have a common border, and if major cities are not very distant from the border, the fear of a pre-emptive attack could still be more important and the incentive to strike first, more attractive.

Political factors are far more crucial than technical conditions, as important as the latter may be. First of all, deterrence supposes a comparison between the importance of the issue and the risk incurred. In 1962, the number of weapons in the US and the USSR was very high and a nuclear confrontation would have resulted in their mutual destruction. At a lower level, the risk of unacceptable losses can be sufficient to prevent a conflict. But the notion of unacceptable loss varies considerably from one country to another and from one period to the next.[286] The reactions of the American and European opinions in case of a terrorist action show that unacceptable losses are not identical for the British, the French or the Spanish, and for the American citizens. Similarly, it seems that in 1999, public opinions are not prepared, in the UK, in Germany or in France, to accept the huge number of casualties which they suffered stoically during the first or the second world wars. However, in case of a crisis, governments must assess the psychological impact of their retaliatory strikes not after an accurate knowledge of what the enemy considers as unbearable, but after their own guess of what the opponent deems acceptable.

It must be added that, in 1962, both governments were led by rational calculations. Few people doubt that, if nuclear weapons had been available to him, Hitler would have used them, whatever the cost for his country. It could be assumed that at present, a couple of dictators would have a similar reaction. Moreover, it is very difficult

---

[286] Beatrice Heuser, *Nuclear Mentalities*, MacMillan, London, 1998.

to assess what might happen if a government was pressed by extremist religious groups, which would not be prepared to accept that the triumph of their faith does not deserve the death of a huge number of people, or even their own destruction.

Second, the Cuban missile crisis was concluded peacefully, because the issue was not as important for the Soviets as for the Americans. For Moscow, the purpose was only to protect Cuba from a possible invasion, and to avoid the repetition of the Bay of Pigs incident. They were not prepared to sacrifice the USSR to save Castro. For the US, it was a vital issue, since Soviet missiles based less than 2 000 km from the Florida shore could reach their targets in a couple of minutes, before a decision could be made to retaliate, to protect the territory or destroy the aggressor. Khrushchev was able to retreat, because the issue was not vital for his country; Kennedy could not step back. Fortunately, there has never been, during the cold war, a crisis in which both parties would consider the issue as equally critical for them.

Third, in 1962, China had not yet tested its first weapon, London and Paris supported immediately the US, so that the crisis was managed by only two parties. Nobody can be sure that deterrence would have prevailed if a third country, China for that matter, had been in a position to raise the stake and had pressed Moscow to refuse the ultimatum issued by Kennedy.

Fourth, deterrence supposes that each party accepts the status quo. After 1962, neither side has attempted to change the balance of forces in areas of the world protected by the nuclear umbrella of his opponent. But both countries tried to find new weapons which could break the stalemate. The precision of the missiles was dramatically improved to try and destroy preventively the silos of the opponent. Anti-missile systems were set up and, finally, the two superpowers decided in 1972 to limit the number of Anti Ballistic Missiles (ABM) which they were authorised to deploy. The number of missiles and of nuclear devices was increased to unbelievable levels in an attempt to outnumber the enemy. The intermediate-range missiles were deployed to prepare a war in Europe which would not threaten the territory of the two superpowers. The Strategic Defense Initiative (SDI) programme was launched to shield the US territory from any missile. It is not sure that nuclear arsenals would have been used as instruments of deterrence if a breakthrough had occurred in one of those fields. It must be added that nuclear weapons might have been used by accident, or by mistake, if one of the governments in nuclear weapons states had misunderstood a movement or a decision made by another.

Nuclear weapons are the best instruments of a deterrence policy. For that reason, the "no first use policy" is very dangerous, since it does not allow to use the threat of nuclear weapons as a last resort tool in the management of a crisis. If that commitment was made by all countries, nuclear weapons could normally not be used at all, but if one of enemies does not respect his obligations, he may gain a decisive advantage over the others. As long as they have not been totally eliminated, nuclear weapons should be permitted to play the role which they can have in preventing a war.

Nuclear weapons, however, are not synonymous with deterrence. In a crisis, decisions are taken by policy makers, not dictated by the very nature of the weapons. If governments make an error, if they misunderstand their opponent, if they underestimate the importance of the issue, or if they are persuaded that they cannot afford a humiliating retreat, they could use their weapons as an instrument of annihilation.

If the number of countries capable of initiating a nuclear conflict increased, the risk that nuclear weapons might be used to destroy an enemy, or to blackmail the neighbourhood or the international community, would increase proportionally. For that reason, non-proliferation is a crucial condition for the survival of our civilisation.

1. Genesis and Development of the Non-Proliferation Policy

From 1945 to 1954, the policy to prevent the dissemination of nuclear weapons was based on secrecy. The US, being at that time the only country capable of producing nuclear energy for civilian or military purposes, decided not to provide any other country with technical data on atomic energy.[287] After the Soviet Union manufactured A and H explosive devices[288] (in 1953) the policy of secrecy was abandoned. Since then, it is assumed that - except for a very small number of technical details - any country can acquire the knowledge necessary to produce an explosive device. The key element in non-proliferation is no longer the availability of scientific or technical information, but the possibility to produce fissile materials.

Between 1954 and 1968, international co-operation was developed to expand the use of nuclear energy for peaceful purposes, and non-proliferation became a second rank concern. However, many countries tried not to assist other states in developing a military programme. In particular, most exporting countries - as well as the International Atomic Energy Agency (IAEA) - did not agree to supply nuclear equipment or materials if the recipient country did not promise to use them for peaceful purposes only, under international safeguards. But importing countries remained free to use their domestic resources for military activities.

In 1962, the missile crisis in Cuba emphasised dramatically the necessity to set up a much more efficient policy. This resulted in the Non-Proliferation Treaty (NPT), in which non-proliferation is based on a double mechanism. On the political level, all non-nuclear weapons states, when they sign the treaty, commit themselves not to acquire or attempt to acquire nuclear weapons. Such a commitment supposes that they consider their security better assured if their neighbours do not possess nuclear weapons, even if this means that they will not get nuclear weapons either. On the technical level, the Treaty provides that when a non-nuclear weapons state avails itself of a quantity of nuclear material sufficient to permit the manufacture of an explosive device, the IAEA must verify that the materials are used only for non-explosive purposes. Since 1968, the conviction that security is best assured if one's neighbours do not possess nuclear weapons, and the efficiency of IAEA safeguards on fissile materials are the two pillars of the non-proliferation system.

The period 1968-1995 was used to persuade all countries to adhere to the NPT. That effort met with difficulties and drawbacks (in particular, Israel having launched a military programme in the fifties continued to develop its arsenal; India exploded a device in 1974; Pakistan started its programme in the late seventies; two signatories of

[287] See Bertrand Goldschmidt, *Le complexe atomique*, Fayard, Paris, 1980.

[288] "A" for "atomic" (fission) devices, and "H" for "hydrogen" (fusion) devices.

the Treaty, Iraq and North Korea[289], violated it at the beginning of the nineties). However, when the signatories of the treaty met in New York in 1995, twenty-five years after its entry into force, to decide on its extension for a limited period of time or indefinitely, they were 178 and they decided without voting to extend it indefinitely. At present, they are 187, after Brazil (being one of the few countries which had not yet acceded to the Treaty) ratified its adhesion on 13 July 1998.

2. The Most Critical Threats to Non-Proliferation

The almost universal adherence to the Treaty was a major success for the whole international community, a triumph of reason over blind nationalist ambitions. It did not mean however that the threat had totally and definitively vanished. Several risks are still looming.[290]

2.1 Countries Non-Signatories of the NPT

Four countries still remain outside the Treaty. The first one is Cuba which cannot be considered as capable of producing the components of a nuclear explosive device in the near future, and therefore does not represent a real risk at present. In Israel, a reactor, capable of producing significant quantities of plutonium and a reprocessing plant were built by France in the second half of the fifties, in the aftermath of the Suez crisis. More modern and larger facilities have been completed later, and Israel is generally considered as having about one hundred weapons available, some of them at least being fairly sophisticated. Although a regrettable exception in the universality of the NPT, Israel does not seem to be the most dangerous hazard, since its conventional forces have achieved such a superiority upon its neighbours, and the country enjoys such an unconditional support from the US, that it seems rather unlikely to make use of its nuclear capabilities, even if neighbouring countries were prepared to resume hostilities.

Such is not the case for the other two countries which did not adhere to the Treaty, India and Pakistan. In the mid-fifties, India acquired from Canada a 40 megawatt (MW) heavy water reactor, the so-called NRX, capable of producing sizeable quantities of plutonium. Meanwhile, the Indians built a reprocessing plant. Using the plutonium they had thus acquired, India tested an explosive device in 1974.[291] Since that time, another reactor (called Druva) was built, based on the same technique. In 1993, the president of the Indian Atomic Energy Commission officially disclosed that his country had begun to operate a centrifuge plant in the state of Karnataka, so that India has available both fissile materials, highly enriched uranium (HEU) and plutonium. The country has also tested a

---

[289] Officially, the "Democratic People's Republic of Korea" (DPRK).

[290] Georges Le Guelte, *Histoire de la menace nucléaire*, Hachette, Paris, 1997.

[291] Leonard S. Spector and Mark McDonough, *Tracking Nuclear Proliferation*, Carnegie Endowment, Washington, 1995.

1500 km missile (called Prithvi) and tested 5 more nuclear devices on 11 and 13 May 1998.

In Pakistan, a uranium enrichment facility, based on centrifuges derived from those used in the Dutch facility at Almelo (in the late seventies, a Pakistani engineer, under a fellowship at Almelo, "borrowed" the documents concerning the design of the plant as well as a list of the major suppliers. Back home, he was appointed Director of an enrichment plant). In 1985, the facility started enriching to over 20 % in spite of assurances given by the Pakistani president to Ronald Reagan. Since that time, Pakistan produces fissile materials, which can be used for explosive purposes, in a country where no reactor uses enriched uranium and which has never exported enriched uranium.[292] More recently, Pakistan has built a heavy water reactor at Kushab, but it is not clear whether the Pakistanis have received from China a sufficient quantity of heavy water to operate it at full power.[293] In March 1998, Pakistan tested a 1500 km missile, named Ghauri, based on the Nodong North Korean technology. After stating for years that they were capable of manufacturing nuclear weapons, but did not want to, the Pakistanis exploded at least two devices on 28 and 30 May 1998.

It should be stressed that, from a legal point of view, India and Pakistan (as well as Israel) have consistently refused to sign the NPT. They are thus not committed to refrain from acquiring nuclear weapons. Since they use nuclear materials produced on their soil, and facilities built from domestic resources (which are therefore not under IAEA safeguards) to produce their fissile materials, their military activities are not in breach of any international obligation. However, the NPT considers as nuclear weapons states only those countries which have tested an explosive device before January 1, 1967.[294] Neither India, Israel or Pakistan meet that condition. It would be very dangerous to consider them as nuclear weapons states, since it would mean that a country which did not explode a weapon before 1967 can nevertheless be recognised as such, and that the NPT applies only until it is violated. Moreover, amending one word in the Treaty would unravel it.

From a political point of view, their situation is very dangerous for themselves, for their neighbours, and for the whole world. India and Pakistan have no deterrence policy or strategy. They are antagonised by an old and basic conflict in Kashmir, aggravated by the presence of fanatic religious groups which have already provoked a number of casualties, and their troops are directly fighting each other. The perception of the danger can serve as a crude deterrence and increase the level of tension which can be accepted without resorting to actual war. But if a local conflict gets out of hand, one or the other

---

[292] Recently, it was stated that the US obtained, in 1991, that the level of enrichment at the Kahuta site would not exceed 20 %, and that the commitment had been verified by intelligence agencies. However, a senior representative of Pakistan claimed that the plant never stopped producing highly enriched uranium (HEU). Depending on which statement is accurate, Pakistan had available either 100 kg of HEU (a large part of which has probably been used if the Pakistanis have actually exploded five devices in May) or 500 kg.

[293] Some research has been made on reprocessing technology, but it seems that, for the time being, there is no reprocessing plant in operation in Pakistan.

[294] The nuclear weapons states are, in chronological order of acquisition: the US, the USSR (at present Russia), the UK, France and China.

government can be tempted to use its nuclear weapons from fear of a pre-emptive strike from the other, and that temptation is not matched by the certainty of a deadly retaliation. In addition, large cities and vital centres are not too distant from the border, and can be reached in a couple of minutes once one of the enemies can make use of intermediate range missiles. That circumstance reduces dramatically the risk of retaliation.[295] Instead of facilitating the management of a crisis (as was the case between the US and the USSR after 1962), nuclear weapons may transform a local incident between India and Pakistan into a disaster.

The five nuclear weapons states can hardly underline the risk created by the arsenals in India and Pakistan, since they would be suspected to point their finger to others in order to distract the attention from their own situation. One can only deplore the silence observed by all countries and all non-governmental organisations (NGOs) which have, in the past, devoted so much of their energy, and which are still using their talents, to mobilise the international public opinion against the risks resulting from the nuclear arsenals of the five nuclear weapons states. At present, their arsenals are by far less dangerous than those of India and Pakistan. It may be that the influence of NGOs is by far larger in western democracies, and their task much more difficult in other countries. But the reasons for the weakness of NGOs in India, Israel or Pakistan are not a valid excuse for the attitude of those non nuclear weapons states which continue to consider the nuclear disarmament of the five nuclear weapons countries as their only target, and have totally ignored what is by far the most dangerous threat for peace in the world since the end of the cold war.

2.2 Secret Facilities

The second most important hazard is the risk of clandestine activities. Since the mid-fifties, when the peaceful applications of nuclear energy were developed, all experts agreed that nuclear facilities could not be built secretly, because they were too large and too characteristic, and they would be detected immediately. It was felt also that a large number of people would be involved in, or informed of, a military programme, and that it would meet some opposition from internal political opinion, forbidding its concealed implementation. In 1991, after the end of the Gulf war, the joint action of the IAEA and the United Nations Special Commission (UNSCOM) has uncovered a huge secret programme in Iraq, which had until then fooled not only the IAEA inspectors, but also the best intelligence agencies in the world. It is at present considered that the programme included at least thirty sites (only one had been declared by the Iraqis), some 10 000 persons were involved, and the order of magnitude of the budget (based on

[295] It should be stressed that, in 1962, the US threatened to use their nuclear weapons because the USSR was installing in Cuba intermediate range missiles, which could reach their targets in the US in two or three minutes (not in half an hour, which would be required if they were launched from the Soviet territory). The risk that they might be destroyed before being able to retaliate was considered as unacceptable by Washington, since the USSR would be tempted to use its weapons and not consider them as a deterrence instrument.

western costs) was in the range of US$ 10 billion.[296] A few months later, as soon as the IAEA could apply its safeguards system in the country, the inspectors demonstrated that North Korea was trying to conceal some of its activities, in violation of the obligations resulting from the NPT.

Thus, within one year, the common wisdom of all nuclear experts over the past two decades proved to be wrong. It was possible to carry out clandestinely a huge military programme, in particular in countries where there is no freedom of information, which are almost closed to foreigners, and which are governed by a brutal regime. In such cases, when the political pillar on which the NPT is based disappears, namely when the country is not prepared to abide by its commitment not to acquire nuclear weapons, the efficiency of the non-proliferation policy rests only on the second pillar, the technical difficulty of obtaining the quantity of fissile materials necessary to manufacture an explosive device.[297] Iraq and North Korea seem to have been stopped in the process of producing fissile materials before they had gathered a sufficient quantity. But in the future, a major effort must be made by the international community to reduce the risk of clandestine activities, or at least to increase considerably the capabilities to detect in time a concealed military programme. Otherwise, the risk of violation of the NPT could materialise. In particular, some people consider that Iraq is prepared to resume its clandestine activities as soon as the vigilance of the international community has been reduced. Since August 5 1998, Iraq has refused to cooperate with UNSCOM, and since 13 September, they have opposed all international inspections on their territory. So far, there has been no reaction from the UN. After the air strikes carried out by the US and the UK in December 1998, Iraq has refused the resumption of inspections by UNSCOM. Since that date, the members of the Security Council have discussed the possibility to monitor only the facilities identified by UNSCOM as having been used previously for the secret Iraqi programme, and to lift the economic sanctions.

2.3 Illegal Trafficking of Fissile Materials

The more difficult it is to produce fissile materials domestically without alerting an international safeguards system, the more pressing the incentive should be for a proliferator to acquire weapons grade materials on the black market, if such a market exists. Until 1990, stringent measures of physical protection had been taken in all countries in which sizeable quantities of nuclear materials were handled. However, after the collapse of the USSR, fears were raised that fissile materials could be stolen

---

[296] Michel Saint Mleux and Georges Le Guelte, "Les armes de destruction massive en Irak", *La revue internationale et stratégique*, n° 31, autumn 1998.

[297] A group of international experts, created by the United Nations in 1967, considered that a "significant quantity" of fissile material is the amount which would be necessary, for a country never having tested an explosive device before, to manufacture "a crude weapon". This quantity was declared to be 8 kg for plutonium, and 25 kg for highly enriched uranium (HEU, uranium containing 93 %, or more, of uranium 235). Those figures have been challenged. In particular, it was argued that countries with a long experience in nuclear weapons can produce a device with smaller quantities. The order of magnitude, however, is valid.

(particularly in reprocessing plants and storage facilities where weapons grade plutonium was handled for military or civilian purposes) and sold to possible proliferators. These fears were not unreasonable, because the surveillance mechanism in place before 1991 could not be swiftly and easily replaced by a new system, based on techniques used in western countries. Apprehension was expressed, also in Russia, because the personnel responsible for the materials were not paid, and the incentive might therefore be strong to look for alternative sources of income.

It seems that illicit trafficking has so far taken place primarily on radioactive materials which are of no use to manufacture an explosive device. A small number of cases have been reported in which people were arrested trying to sell weapons grade materials illegally acquired, but the quantities involved were far too small to be significant. Dual use equipment were also seized, before reaching their final destination, but trafficking of important components has not so far been seriously confirmed.

Important efforts have been made in particular by the US, the European Community and some of its member states (mainly Sweden and the UK), to strengthen all measures of security in the former Soviet Union, especially in Russia. It seems also that the most important quantity of plutonium (30 ton), which had not been assigned to defence purposes, has recently been transferred for storage to the Mayak facility, where physical security has been strengthened.

2.4 Fissile Materials from Dismantled Weapons

A slightly different problem has emerged with respect to the fissile materials made available as a consequence of several disarmament agreements: the Intermediate-range Nuclear Forces (INF) Treaty, and the Strategic Arms Reduction Treaties (START I and II). As the strategic missiles are destroyed, those materials could no longer be used in a conflict between the US and Russia, but as long as they have not been disposed of, they might be utilised to manufacture new weapons (e.g. against other countries), and they risk to be stolen and sold to proliferators or to terrorist groups. The quantities involved are impressive (the figures for Russia are above 500ton. of highly enriched uranium, and some 200 ton. of plutonium, and roughly half of these amounts for the US). By definition, they could serve to produce as many weapons as will be dismantled, i.e. 20 000 to 30 000.

Highly enriched uranium can be diluted with depleted or natural uranium, and the mixture obtained can be used in power reactors to produce electricity. Once the enrichment rate has been reduced to about 3 %, the uranium can no longer be used for explosive devices.

Very different is the situation of plutonium (see Chapter 13, by Goldschmidt, and Chapter 14, by Garwin, for an extensive description of the plutonium case). At present, it is stored in heavily guarded facilities. Whereas the precautions taken are very serious, however, storage cannot conceivably be maintained over the whole half-life of plutonium of 24 000 years. As long as it is not rendered impossible to use it in explosive devices, the plutonium resulting from the dismantlement of US and Russian arsenals will remain a hazard for world security. Currently, there seem to be only two techniques available to make plutonium inoffensive in this regard: vitrification and disposition in geological sites, on the one hand, and the use as Mixed Oxide (MOX) fuel elements in light or heavy water reactors, or in fast neutrons reactors, on the other hand.

Those solutions have both their disadvantages. Vitrified plutonium could be recovered, although the recovery would be long and costly. It is important to underline that the Russians claim that the plutonium is for them an invaluable source of energy, and they do not want to waste it. As for the use in light water reactors, it would not permit to burn all the plutonium loaded. Irradiation in the reactors would affect the isotopic composition of the remaining plutonium and render it less convenient for the production of a weapon, but it would not foreclose its use to manufacture an explosive device. As long as it is contained in the fuel elements, the plutonium is not available, and for some period of time the level of radioactivity is such that it can be recovered only in a sophisticated reprocessing plant. However, after a few decades it would become much easier to recover the elements and to separate the plutonium.

In other words, vitrification is not a definitive solution. Consumption eliminates only the amount of plutonium actually burnt in reactors. However, both techniques would extend considerably the time span between the decision to reuse the fissile materials in nuclear explosive devices and the moment when the weapons would be operational. Both techniques would take time (the only project under consideration to burn plutonium via MOX in four Russian reactors, presently in operation, would burn 1.6 ton per year). They would be very costly. For example, to use more quickly the 200 ton of plutonium in Russia, it would be necessary to build a number of reactors and fuel fabrication plants, in addition to those in operation. Using all Russian plutonium, deriving from the weapons dismantlement, in fast breeder reactors (where it would be almost inaccessible to a nationalist regime or to traffickers, and where it could be burnt at a rate consistent with the Russian energy requirements), would necessitate the construction of about ten 1 000 megawatt electric (MWe) fast neutrons reactors, the cost being in the order of at least US$ 50 to 60 billion.

In September 1998, 5 years after the signature of START II, the agreement has not yet been ratified by the Russian Parliament. The American and Russian Presidents have agreed in September 1998 that each country would "neutralise" 50 ton of plutonium deriving from its arsenal. However, as no solution has been found to use the plutonium as a fuel or to vitrify it, a committee set up by the so-called "G8" countries has apparently reported, in June, that the only possibility was to store it. All the plutonium recovered from the dismantlement of the weapons will therefore remain in a form which would permit to reuse it rapidly to produce new explosive devices.

In devising a policy to render military plutonium inaccessible for the production of explosive devices, one of the most important obstacles is that anti-nuclear movements oppose violently the vitrification and disposition of nuclear wastes, and they oppose the construction of new reactors. Not a single politician subject to an election can be expected to make the effort of persuading voters that they should pay taxes to dispose of plutonium in a foreign country if that means losing votes.

It must be stressed that it is not possible to refuse all solutions which would preclude the re-use of military plutonium for explosive purposes, while advocating nuclear disarmament. A didactic effort is needed to convince public opinion that, after the cold war, the most important question, in this context, is no longer whether coal, wind, or oil should be used to produce electricity rather than nuclear energy. The main problem is to decide if, for the generations to come, the risks resulting from vitrification or from the operation of nuclear reactors are more serious than the indefinite storage of 300 ton of plutonium which could, in a few weeks, be converted into tens of thousands of nuclear weapons.

In spite of all those very difficult problems, the situation concerning the risk of proliferation of nuclear weapons is, by far, less gloomy than it has ever been over the last fifty years. Almost all non-nuclear weapons states have adhered to the NPT, and are therefore committed not to produce or acquire any nuclear explosive device, the security of fissile materials in Russia has been improved, and the possibility to produce or separate fissile materials clandestinely has been taken into consideration. However, three problems remain to be solved : the risk resulting from the Indian, Israeli and Pakistani arsenals, the difficulty to detect in time secret facilities to produce fissile materials, and the risk of illicit trafficking of materials recovered after the dismantlement of nuclear weapons in Russia.

In April 1998, the meeting of the Committee preparatory to the NPT Conference in 2000 clearly emphasised that no further progress can be made as long as it remains impossible to mention the existence of the Israeli arsenal and still more to discuss it, or to consider its possible elimination. A large number of countries will not accept to say anything against the Indian and Pakistani weapons, and even to support any action against Iraq, Iran or North Korea, since they perceive the indulgence demonstrated to Israel as an unacceptable discrimination. That problem must be considered very seriously, because non-proliferation must be dealt with as a whole. A country can be persuaded that its security is more properly assured if none of its neighbours can acquire nuclear weapons. But if they do not believe that a breach of commitment would be immediately detected and sanctioned, the political confidence underlying the NPT might be questioned. If North Korea had been in a position to actually manufacture a weapon, would South Korea and Japan have indefinitely considered that their security was best assured if they had no nuclear military programme?

3. The Safeguards System: a Barrier against Proliferation

The nuclear non-proliferation policy is one of the exceptional areas in which an international safeguards system has been set up to verify that the obligations taken are actually respected. Until 1968 and the signature of the NPT, the IAEA safeguards system applied only to materials or equipment supplied by the Agency itself, or on which the supplier had demanded such guarantees, or for which the safeguarded country had requested an intervention of the Agency (which was exceptional).[298] The mechanism became of wider application after the NPT was concluded, since its article III provides

[298] David Fischer and Paul Szasz, *Safeguarding the Atom*, Stockholm International Peace Research Institute (SIPRI), 1985.

that all non-nuclear weapons states signatory to the NPT should subject their nuclear activities to IAEA inspections. As the number of signatories increased over the years, the safeguards system became also of almost universal application.

3.1 The Instruments of the Safeguards System

The IAEA safeguards system is based on a detailed and accurate accountancy of all nuclear materials (uranium, thorium and plutonium) used in the civilian activities of a country.[299] The safeguards system assumes that if the IAEA finds at the outlet all the materials introduced in the civilian nuclear industry, it is possible to conclude that there has been no diversion to unauthorised purposes.

The verification process consists of three successive phases:

- All operators must send to the Agency the technical data of facilities handling nuclear materials. The inspectors can thus check that in designing a plant, no particular mechanism has been inserted to facilitate the possible diversion of nuclear materials. Basically, they can decide on the safeguards strategy for each specific plant; they can identify which areas deserve a more accurate surveillance, which operations are especially sensitive, or at which places it will be easier and more important to make some measurements.
- For each movement of nuclear materials in the plant, the operator must immediately inform the Agency and supply a copy of the accountancy document, so that the inspectors can trace all nuclear materials and keep an up-dated accountancy of all materials in the facility.
- Periodically, inspectors perform a physical verification, first checking that their accounting data and those kept by the operator are identical. Then, they check the materials themselves, quantitatively and qualitatively, by measuring and taking samples which will be analysed both by the operator and in the Agency's laboratory, to make sure that the chemical, but chiefly the isotopic composition of the material are those which have been declared by the operator.

3.2 Technical Limitations of the Safeguards System

There are limitations, and shortcomings in this mechanism, some of them are technical, but the most important loopholes are of legal nature.

Technically, it is impossible to prove that an event did not occur, and the Agency cannot demonstrate scientifically that there has been no diversion of nuclear materials. In practice, its conclusions are very prudent, and the Agency only states that inspectors did not find evidence of a possible diversion. This is a very honest and clever statement, but it may not always be easily accepted by the public, which prefers absolute certainties, and has some difficulty in understanding that it is physically impossible to achieve a more

---

[299] David Fischer, *History of the International Atomic Energy Agency: The First Forty Years*, IAEA, Vienna, 1997.

accurate conclusion than a very high probability that nothing illegal or dangerous has taken place.

It should be added that, when considering the basic features of a plant, the inspectors must determine the inevitable losses associated with each facility, depending on the nature of its activities, and on the particular characteristics of the plant. For a storage facility or a reactor, such losses are null or exceptional. But it is not the case in a reprocessing or enrichment facility, or in a fuel fabrication plant. In all those facilities, nuclear materials are at some moment in a liquid form, and at another in a metal form. They can be shaped by tools which will produce scraps, not all of which can be totally recovered. Depending on a number of factors, a percentage of the materials, generally very small, but never null, cannot be recovered. As long as the actual losses do not exceed that percentage, the inspectors can assume that they do not mean a possible theft or diversion; they are normal, unavoidable losses. Basing their judgement on their knowledge of the plant, and on the equipment (such as cameras) which can be placed to observe the movements in critical areas of the plant, they generally reach the conclusion that, in any case, the material missing has not been removed out of the boundaries of the facility.

However, the inspectors cannot materially measure the materials, and public opinion has sometimes difficulties in understanding that they do not have any suspicion as to the fate of the materials, since they possess no scientific and measurable evidence of their situation. Such misunderstandings can evolve into political incidents, as was the case in 1995, when a newspaper reported that 70 kg of plutonium was missing in a Japanese facility. In fact, plutonium had accumulated on the floor, the walls and the ceiling of cells in a plant manufacturing fuel elements for a fast neutron reactor. The situation was very well known, the whole amount of plutonium was accurately recorded in the accountancy of the facility, all data were regularly sent to the Agency, and a special device had even been installed to measure the plutonium in the cells. Not a single gram of plutonium had therefore been diverted for use other than the fuel elements for the reactor, but the journalists had not realised the difference between the measurement by the inspectors of a lack of material, and the risk of diversion. It is not clear that the newspaper in question did ever acknowledge its own mistake and explain its readers the details of the false rumours it had spread out, although they were very embarrassing for the operator and for the country where the incident took place.

It should be added that the safeguards system does not pretend to identify the smallest possible diversion. In Iraq, in the eighties, a few pins were manufactured using uranium produced domestically. They were irradiated clandestinely in a reactor built by the USSR in the sixties, which was under IAEA safeguards. Finally, five to six grams of plutonium were separated secretly. In Rumania, after Ceausescu had been overthrown, the Government reported that a few milligrams of plutonium had been produced and separated in similar conditions.

The IAEA is aware of these shortcomings, but detecting all possible diversions would require a much larger budget and a far greater number of inspectors. In fact, the target of the system is to detect the diversion of a significant quantity of nuclear materials in a time-frame shorter than the time necessary to turn the material into a nuclear explosive device. The significant quantity has been defined as 8 kg for plutonium and

25 kg for highly enriched uranium. As for the period of time necessary to manufacture a device, it depends very much on the chemical form, the physical form, and primarily on the isotopic content of the material. This will determine the frequency of the inspections. For highly enriched uranium or plutonium, if the quantity involved is greater than a significant quantity, the frequency of the inspections will be very high. For natural or depleted uranium, the inspectors may come once per year, even if the total amount of uranium 235 contained is above 25 kg. Although this is very reasonable, the diversion of very small amounts of materials may therefore remain undetected.

3.3 Legal Limitations of the Safeguards System

The technical features of the safeguards system are permanently improved, and if some limitations remain, they do not jeopardise its overall efficiency. The major loopholes lie in the legal characteristics.[300]

The IAEA is not a signatory of the NPT, because when the NPT was signed some of the IAEA member states (including Argentina, Brazil, China, France, India, Israel, Pakistan, etc.), which did not adhere to the NPT, stressed that the agreed role of the IAEA was only to discharge the responsibilities defined by the Agency's Statute. When the provisions of a treaty signed by some of its member states does not contradict the Statute, the IAEA can combine both obligations, but they can not prevail over the Statute. Therefore, when a country adheres to the NPT, its obligations to the Agency are not defined by the Treaty, but by a bilateral agreement, which must be concluded and ratified according to the procedures in force in each country. This was for example the reason why the IAEA could not begin full scope safeguards in North Korea before April 1992, although that country had ratified its adhesion to the NPT on 12 December 1985.

Until 1968, the IAEA had been requested to safeguard only nuclear materials or facilities which it had supplied to a particular country, or which had been exported by a supplier demanding the application of international safeguards. These safeguards were implemented in accordance with a standard agreement fitting the surveillance of a given quantity of materials or of a specific facility.[301] When Article III of the NPT made it an obligation for all non-nuclear weapons states to place all their activities under safeguards, it became necessary to define a new type of agreement, adapted to the surveillance of all nuclear activities in a country.

Under pressure from the major industrial states, it was decided that the new type of agreement would be defined, not by the Secretariat of the Agency (which was the proper solution if the agreement was to outline the technical solutions necessary to implement the legal obligations contained in the Treaty), but by an open-ended committee of the Agency's member states. This meant that not only technical matters, but also political decisions had to be made. In fact, Article III of the NPT provides that safeguards should be applied "*for the exclusive purpose of the fulfilment of ... obligations assumed under this Treaty with a view to preventing diversion of nuclear energy from*

[300] Georges Le Guelte, *op. cit.*

[301] This standard agreement was the so-called "INFCIRC 66/Rev. 2".

peaceful uses to nuclear weapons or other nuclear explosive devices". This is an ambiguous wording, since it is not very clear if the purpose of the safeguards system is to make sure that the country will "... *not manufacture or otherwise acquire nuclear weapons or other explosive devices* ...", as stated in Article II, or if its only objective is to check that the acquisition of nuclear weapons is not obtained through diversion from peaceful to military activities. In other words, it does not say clearly if safeguards must apply to any activity in the country and must verify that there are no clandestine activities, or if they should be limited to verifying that all activities declared civilian by the state are not diverted to any other purpose, but not try and check the absence of undeclared facilities.

The standard agreement adopted in 1971[302] uses a different language, but it contains a basic commitment as ambiguous as that in Article III of the Treaty. The text does not explicitly specify if the search for possible clandestine activities is authorised or not. It provides that safeguards should apply to "*all source or special fissile material in all peaceful nuclear activities within the territory of the state, under its jurisdiction or carried out under its control anywhere* ...", but "... *for the exclusive purpose of verifying that such material is not diverted to nuclear weapons or other nuclear explosive devices*". However, while all the rest of the agreement contains a very detailed and sometimes minute description of the modalities of safeguards on declared materials or facilities, not a single word is devoted to specifying the techniques or the instruments for the detection of concealed activities.

In practice, all nuclear experts of the time were convinced that a nuclear facility could not materially be built and operated clandestinely, since its size and characteristics would not allow it to be hidden. In addition, nuclear industrialists did not want to be bothered by international inspectors, and when some IAEA inspectors tried to go beyond what they were strictly permitted, they were reported to their superiors which did not facilitate their promotion in the hierarchy of the Agency. Thus, although the text of the agreement does not preclude the search of possible clandestine activities, all safeguards experts inside and outside the Agency professed, as a Bible, that the purpose of the Agency's safeguards system was only to verify that declared materials in declared facilities were not diverted from peaceful to military purposes. This became by far the most important limitation in the machinery.

When the Agency reaches the conclusion that a significant quantity of nuclear material has been diverted, it can only inform the UN Security Council and General Assembly. It is then up to the international community to decide on measures to be implemented. There is obviously no other conceivable solution. However, it means that the reaction of the international community is necessarily a political decision, based on the balance of forces at a given time. For example, in the case of North Korea, the "agreed framework" of October 1994 provides that two light water reactors (each of 1 000 MWe) will be built in North Korea free of charge, and that 500 000 ton of oil will be delivered each year. Some people considered this as a substantial reward for violating the Treaty. In exchange, the North Koreans accepted to freeze their nuclear programme under IAEA safeguards. This means that, until the two Pressurised Water Reactors

[302] This standard agreement is the so-called "INFCIRC/153".

(PWRs) are completed, the IAEA can at best make sure that the provisions of the agreement reached between the US and North Korea are implemented. The Agency cannot verify that the provisions of the NPT are respected. For example, the 1994 agreement does not permit the IAEA to visit the sites which could provide some information on its past activities, nor the site recently identified by satellites and supposed to be an underground cavity where it is assumed that another reactor might be clandestinely built in the future.

3.4 Improvements in IAEA Safeguards: Strengthened ("93+2") Safeguards System

The discovery of a huge clandestine nuclear programme in Iraq, after the Gulf War, underlined that the most important shortcoming of the IAEA safeguards system was the impossibility for the inspectors to try to identify possible clandestine activities in a country. In the beginning of 1991, the IAEA Board of Governors endeavoured to remedy that loophole to the largest extent possible. An attempt was made first to modify a few legal provisions in the standard agreement which had been used by the Iraqi diplomacy to try to deny any wrongdoing in their actions. The consequences of the decisions made by the Board were very limited, and would not increase significantly the competence of the inspectors. A second attempt was launched in 1993, intended to be approved by the NPT extension Conference in 1995, which for that reason was frequently referred to by the name "93+2". Its declared purpose was to provide the IAEA with a safeguards system capable of detecting anomalies which might indicate the existence of possible clandestine activities, which would meanwhile remain as efficient as the existing one to detect in time the possible diversion of declared nuclear material from peaceful activities to the production of an explosive device.

3.5 Preliminary Legal Amendments

After the Gulf War, the United Nations Security Council created UNSCOM to identify in Iraq the facilities used to manufacture weapons of mass destruction. UNSCOM was authorised to use some exceptional methods, not available to an international organisation in normal circumstances. With the assistance of UNSCOM, the IAEA was able to make a detailed inventory of facilities which had so far remained unknown, and of nuclear materials which had not been reported.

In almost all cases, the Iraqi authorities claimed that they had not violated the provisions of their agreement with the Agency. For example, several metric tons of natural uranium extracted from a phosphate mine had never been declared to the Agency. But the Iraqis explained that, according to their agreement, natural uranium produced domestically had to be declared only when it was in a form appropriate for enrichment (UF_6 for example) or when it could be used in a reactor (as a metal or as UO_2). There was no such obligation as long as it was in the form of concentrates of U_3O_8. Similarly, facilities under construction were discovered for the enrichment of uranium in the isotope uranium 235 by centrifugation and electromagnetic systems. They had not been declared either, but the Iraqis maintained that, according to their agreement, a new facility had to be declared to the Agency 180 days before the nuclear materials were introduced. As they needed more than 180 days before introducing the materials, they were not guilty of any breach of their obligations.

All the claims made by the Iraqis were valid, the loose wording had been imposed in 1971 by the most important producers of uranium or by industrial countries which wanted to protect their nuclear industry against the burden of international safeguards. They provided Iraq with legalistic reasons to argue that they had not violated their obligations. And in fact, the main question was to decide whether the purpose of safeguards is to verify that the country does not try to produce a nuclear explosive device, or only to check that declared materials are not diverted to unauthorised uses when they are in declared facilities. To quote the then Director General of the IAEA, Hans Blix, the point was not to decide if Iraq had been guilty of hiding from the Agency a few kilogram of natural uranium in the form of chloride, or the separation of a few grams of plutonium (the only mistakes for which the Iraqis could not find an excuse in the text of the agreement). The main issue was that Iraq had violated the commitment made in signing the NPT by setting up a clandestine nuclear programme composed of some 30 concealed sites, operated by about 10 000 people, and designed to manufacture a nuclear weapon.

However, the use by the Iraqis of the wording of the agreement inclined the Board of Governors to amend the text adopted in 1971, to request the declaration to the Agency of all uranium produced domestically, whatever its form, and to declare a new facility 180 days before the beginning of construction, not before the introduction of nuclear materials. Since some countries were not prepared to accept those provisions if they did not apply to all states, including the nuclear weapons states, it was decided that such information would be transmitted to the Agency on a voluntary basis. Interestingly enough, in the beginning only the member states of the European Union accepted to provide this information. They were quickly followed by a number of others, because it appeared that those refusing could be considered as having something to hide.

Already in December 1992, the Board of Governors (before discussing these amendments) had made a decision which seemed to be much more important. The standard safeguards agreement of 1971 provides that, under very restrictive conditions and in a very limited number of cases, the Agency can perform a special inspection in a facility which is not usually inspected, if this permits to clarify an anomaly found in a safeguarded facility. The Board accepted that a special inspection outside a declared facility could be requested by the Director General, even if the information on that facility had been received from external sources (meaning, in practice, the intelligence agency of another country).

In principle, the decision was very important, because it was recognised that the responsibility of the inspectors was not confined to the accountancy of nuclear materials, that their duty was also to examine possible undeclared facilities, and that they can make use of information received from intelligence sources. All these elements were contrary to the principles so far accepted concerning the purposes of the safeguards system, as well as the instruments available to the inspectors. However, the only occasion when the Agency actually tried and used this possibility was in North Korea, where the authorities refused sternly to let the inspectors into two designated places. Their refusal was the reason why the UN Security Council was involved in the problem. As the inspectors were finally not permitted to enter the two places, some people doubt whether the Agency will ever be in a position to make use of the special inspections in other instances.[303]

The shortcomings in the text of the agreements, corrected by the Board in 1992, were of limited importance to the Agency. Even the absence of these shortcomings would not have given the inspectors the possibility to detect the clandestine activities under way in Iraq. It was therefore felt that further modifications were necessary to strengthen the safeguards system, and make it more capable of identifying clandestine activities. The design and the negotiations of a new system required a much longer time than expected in 1993, and on the eve of the Extension Conference, in 1995, it was clear that not all countries were prepared to accept the modifications proposed by the Director General. Further discussions were necessary. In fact, a new regime was agreed only in May 1997. Its implementation is currently being negotiated with each non-nuclear weapons state signatory of the NPT, since an agreement with each individual country is necessary to give this new regime a legal authority equivalent to that of the safeguards agreements currently implemented.

3.6 The Strengthened ("93+2") Safeguards System

The new system is based on the principle that the inspectors should be informed not only of the nuclear materials and of the facilities declared by the safeguarded countries, but that they should also be informed of all other elements which can contribute to the production of fissile materials or which could be used to manufacture a nuclear explosive device.[304] For example, the IAEA should receive the necessary information on institutions which carry out research on technologies linked to the production of fissile materials (various uranium enrichment or reprocessing techniques, in particular) even if it is confined to theoretical work and if there is no nuclear material actually involved. The Agency should be informed of the status of a nuclear facility which has been shut down. Such a facility is supposed not to contain anymore nuclear materials. Obviously, the Agency can make use of all relevant information which can be found in open literature, in the media, in communications made by scientists in international meetings or seminars, or in the reports of its own inspectors.

The purpose is to give the inspectors the possibility to make in the country under consideration a global assessment of all activities which would be necessary if that country was actually conducting a concealed programme, and to permit the inspectors to compare all information they receive and verify if they are all consistent with the purely peaceful activities which have been declared. For example, if some laboratories are studying very actively various techniques for uranium enrichment, while the country operates only a natural uranium plant, and does not intend to switch to enriched uranium reactors, the Agency may find that a particular attention should be paid to that country.

---

[303] In fact, even up to January 1999, the inspectors have not been able to visit either of the two places.

[304] See "Future safeguards", *IAEA Bulletin*, December 1997.

Or, if arrangements are being made in a country to begin the exploitation of a uranium mine of limited capacity, in which the uranium content is low and its extraction difficult, the Agency may wish to make sure that it will have a complete report of the whole production. This is particularly the case if the country does not operate a reactor and cannot hope to export its uranium on a market where uranium is abundant and cheap, or if the country operates power reactors, but imports all its fuel elements.

In addition to confronting pieces of information from different origins, emanating from the safeguarded state itself or from other sources, the inspectors will be authorised to take samples from the environment, in nuclear facilities, or in places where anyone can have access (such as beaches, parks, or river banks). The samples will be analysed simultaneously by both the interested country and the Agency. If they contain any element not consistent with the declared activities of the country, the discrepancies must be clarified.

It will take time before all those new techniques can be utilised profitably. As time passes by, their efficiency will increase, since the Agency can rely on a larger data base and compare one piece of information with elements previously received from the country safeguarded or with data supplied, for example, by an exporting country. The most difficult challenge for the inspectors is that they must perform simultaneously two different tasks. When verifying that there has been no diversion of declared materials, as they have done so far, their task is basically quantitative. This activity relies heavily on the accuracy of their measurements and on their accountancy. If a significant discrepancy is found, it is an infraction. When they consider the possibility of concealed activities, however, their duty is primarily qualitative, based on the comparison of information. This requires a good knowledge of the whole nuclear fuel cycle and of all equipment or materials necessary to conduct a military program, and sometimes even a good knowledge of other types of industry. The purpose of their work is to verify possible inconsistencies in data from various sources. If an inconsistency is discovered, it is not necessarily an infraction. Simply, it must be checked until the situation is perfectly clear and until they can make sure that it does not signal a clandestine activity.

It must be stressed, that whereas in this respect the action of the Agency may be compared to the role of an intelligence service, the inspectors can - at best - detect an unresolved inconsistency between the declarations of the safeguarded state and the result of their own investigations. They may, perhaps, identify activities carried out in undeclared conditions, but unless they are very lucky, they will be unable to detect the location of the plant where a clandestine activity is performed. Only intelligence agencies may be able to locate secret facilities, and even when the new safeguards system is operating satisfactorily, the Agency will rely heavily on the intelligence system of its member states to identify the places where the inspectors should carry out special inspections. The failure of the best intelligence agencies to detect in time the preparation of the recent nuclear tests in India in 1998, but also the test in 1974, demonstrates that there is room for improvements in this field. Even if the IAEA safeguards system has been strengthened, after the discovery of the Iraqi clandestine programme, the detection of possible concealed activities will remain one of the weakest features of the non-proliferation regime.

Conclusion

The almost universal adherence to the NPT and its indefinite extension were remarkable achievements. They were an unprecedented and unexpected success for the non-proliferation policy. However, the risk of dissemination of nuclear weapons remains the most important hazard for the world. This risk has not disappeared.

Non-proliferation is primarily a political issue. India has shown that in some countries nuclear weapons remain a symbol of power and their acquisition can become a major asset in the domestic politics of a country. It is also true that if almost all countries forego nuclear weapons, and if the safeguards system is sufficient to provide a credible assurance that the commitments made will be strictly respected, a particular state will consider that a general abstention is a far better guarantee of its own security than the possession of its own arsenal. But if the number of nuclear weapons states increases, if the international community proves unable to react energetically, or if the safeguards system is not credible, a number of countries may reconsider their policy.

Since all countries, except India, Israel and Pakistan, are at present signatories of the NPT, those who want to acquire their own nuclear arsenal must withdraw from the Treaty. This might prove to be very risky, since the reaction of the neighbours and of the international community may be very strong. A few countries, which are not open to foreigners and where transparency on governmental activities does not exist, may try either to buy fissile materials on the black market (which will remain impossible if stringent measures are taken), or to build secret facilities to produce the fissile material needed. The new techniques available to the IAEA to detect such concealed facilities will make this more and more difficult.

Finally, such countries could try to divert nuclear materials from their declared peaceful activities to the production of explosive devices. But the IAEA safeguards system makes this extremely difficult. So far, among all countries (including Iraq and North Korea, not to mention India, Israel and Pakistan) which have tried to manufacture their own weapons and which have built dedicated facilities, none of them has ever attempted to divert nuclear materials from a facility under safeguards, because the risk for them to be caught would be too high. On declared materials in facilities declared to be civilian, the surveillance performed by the IAEA is very strict The risk of failure of the system is not nil, but it can be considered as low as a risk may be in any human activity. Of all solutions which a proliferating country can try and use, diversion of fissile materials from a facility under safeguards is by far the most difficult, the least likely to succeed, and the least probable.

Chapter 18 The Risk of Proliferation: the Role of International Agencies

by
David Fischer*

1. Treaty Coverage: The NPT and Regional Weapon-Free-Zones[305]

The international and regional bodies that play a significant role combating nuclear proliferation are:

• the International Atomic Energy Agency (IAEA) established in 1957. Its creation was proposed by US President Eisenhower on 8 December 1953 in a speech to the General Assembly. The IAEA promotes the peaceful use of nuclear energy and seeks to ensure that any nuclear assistance with which it is associated does not further any military purpose. Today, the chief task of the IAEA is to verify compliance with the most important component of the international nuclear non-proliferation regime - the Nuclear Non-proliferation Treaty (NPT) of 1968 - which has been accepted by more than 180 non-nuclear-weapon states and five-nuclear weapon states,

• the regional bodies that service regional treaties prohibiting nuclear weapons or regulating nuclear activities,

• the global bodies that promote international peace and security and seek to enforce compliance with international treaties, principally the United Nations and its Security Council.[306]

Five treaties that outlaw nuclear weapons in the regions they cover are now in force or are awaiting the ratification needed to bring them into force. They are:

- the Antarctic Treaty,
- the Tlatelolco Treaty, covering Latin America and the Caribbean and surrounding seas,
- the Rarotonga Treaty covering the South Pacific,
- the Pelindaba Treaty which, when it enters into force, will cover Africa, and
- the Bangkok Treaty designed to do the same in South East Asia.

---

* David Fischer, a consultant on nuclear non-proliferation and safeguards, has been an Assistant Director General for External Relations at the International Atomic Energy Agency (IAEA) in Vienna.

[305] This survey will deal only with the role of the international (and certain regional) agencies in combating *nuclear* proliferation - the only aspect of the wider subject "international (and regional) agencies and proliferation" that is relevant to the theme of "the prospects of nuclear energy".

[306] To the extent that NATO carries out the peacekeeping initiatives of the Security Council, it may also be said to be in this category, but so far NATO has not played any role in directly inhibiting nuclear proliferation.

The first three treaties cover vast stretches of ocean as well as the land masses of the areas they encompass. It would not be feasible today to verify that the oceans are, in practice, "nuclear-weapon-free". However, with the end of the Cold War and the withdrawal of most sea-borne tactical nuclear weapons, the nuclear-weapon states have less incentive to deploy nuclear weapons in the regions concerned.

For many years, states in the Middle East have sought agreement on a treaty that would ban all weapons of mass destruction from the region. At the insistence of Egypt (and earlier, Iran), the United Nations General Assembly has adopted numerous resolutions recommending the treaty. In recent years, Israel had come out in qualified support. However, since the assassination of Itzhak Rabin, the erosion of the Oslo peace process has made the treaty's prospects - never brilliant - more remote than ever.

The prospects for creating a nuclear-weapon-free zone in Central Asia seem to have improved, but proposals by Belarus to create such a zone in Central Europe are unlikely to bear fruit. There has also been some, so far inconclusive, discussion of a nuclear-weapon-free zone in East or North-East Asia and of an "ASIATOM" (Asian Atomic Energy Agency).

The vast nuclear-weapon-free area that the five existing regional treaties are jointly establishing will cover the entire Southern hemisphere and a sizeable part of the Northern. In this way the treaties open an alternative path to a nuclear-weapon free world. They are still limited in geographical scope but, unlike the NPT, they deny a privileged position to the nuclear-weapon states. The number of states to which the existing regional treaties are expected eventually to embrace is already more than half of those now party to the NPT. They thus provide an increasingly significant option for global nuclear disarmament.

These treaties have already secured significant political gains. At a time when the NPT was still somewhat suspect in the eyes of many of their citizens, they offered Argentina and Brazil a means of ratifying their renunciation of nuclear weapons, they provided South Africa with an additional means - additional to the NPT - of reassuring its fellow Africans that it had turned its back forever on nuclear weapons, and provided for regional as well as international verification of these three ratifications. However, eight crucial states are still missing from the regional agreements; the five acknowledged nuclear-weapon states and India, Israel and Pakistan.[307]

As a rule, the regional treaties rely on the safeguards of the IAEA to verify compliance with their ban on nuclear weapons. Virtually all provide that their member states must ensure that IAEA safeguards are applied to any nuclear items that they export (with some exceptions of exports to nuclear-weapon states).[308] The more recent treaties

---

[307] Moreover, the commitment of at least two non-nuclear-weapon states party to the NPT is, to say the least, dubious: Iraq and North Korea ("the Democratic People's Republic of Korea", DPRK, to give it its official name).

[308] The Tlatelolco Treaty does not address the issue of safeguards on exports but the matter is covered in the safeguards agreements that the IAEA concludes with Tlatelolco parties. The Pelindaba Treaty permits exports of unsafeguarded nuclear material to nuclear-weapon states, probably in recognition of the fact that the main importer of nuclear material from Gabon and Niger is France, and that some of the exported uranium is likely to end up in a French nuclear weapon.

prohibit nuclear exports, unless the importing state has accepted IAEA safeguards on all its nuclear activities - "full-scope safeguards".[309]

The Tlatelolco Treaty contains a protocol in which the nuclear-weapon states undertake to respect the nuclear-weapon-free status of Latin America and the Caribbean. All regional treaties concluded since 1967 have sought similar commitments from the nuclear-weapon states. If a global nuclear disarmament treaty becomes feasible, the category of legitimate "nuclear-weapon states" would obviously disappear and it may be desirable formally to place on record an interpretation for the NPT and the relevant provisions of the regional treaties to reflect this important change (The NPT itself is virtually unamendable).

2. The European Union

The six original states of the European Communities (today the European Union, EU) included a nuclear-weapon state, France. Accordingly, the safeguards applied by the nuclear arm of the Communities, EURATOM (European Atomic Energy Agency), are designed to verify that within the Communities, nuclear materials *are used for the purpose defined by the user* rather than solely for civilian purposes.

EURATOM safeguards are also designed to verify that the EURATOM states are acting in compliance with commitments they have made to other states and international organisations about their use of nuclear material. However, the EU and its predecessors relied heavily on imports of such material, and as a rule the exporting state prescribed that its nuclear exports should be used solely for peaceful use. Moreover, a safeguards agreement concluded in 1973 between the IAEA, EURATOM and the latter's non-nuclear-weapon-states provided that both IAEA and EURATOM would verify compliance with the undertakings those states had made in the NPT not to divert nuclear materials to nuclear weapons.[310]

As a result, in the (currently thirteen) non-nuclear-weapon states of what is now the EU, the IAEA and EURATOM carry out similar verification - or as the former Director General of the IAEA, Hans Blix, put it, EURATOM's non-nuclear weapon states pay twice for the same job. "This may have implications for the verification in the EU states of the obligations that would flow from a cut-off agreement. Means of avoiding the unnecessary duplication of IAEA and EURATOM safeguards have still to be found."

[309] This applies only to exports to non-nuclear-weapon states. To require such safeguards in a nuclear-weapon state would be tantamount to asking it to abandon its nuclear weapons - a requirement that may be desirable but cannot be enforced by the weak lever of nuclear exports.

[310] Published by IAEA as document INFCIRC/193. The Council of Ministers of the Communities approved the agreement with IAEA in 1973, but it still had to ratified by the governments concerned of the EU. For this, and other reasons, it entered into force only in 1975.

3. The Zangger Committee and the NSG

As noted, Article III.2 of the NPT requires its parties to ensure that IAEA safeguards are applied to their nuclear exports to non-nuclear-weapon states. Soon after the entry into force of the NPT in 1970 the main nuclear suppliers already party to the NPT or expecting to be so soon, established the so-called "Zangger" Committee. Its main task was to draw up an internationally agreed list of nuclear materials and equipment the export of which would require the application of IAEA safeguards, thus ensuring that the fierce commercial competition of the 1970s and 1980s would not take the form of offering unsafeguarded nuclear supplies - as to some extent it had in the 1950s and 1960s.[311]

On 18 May 1974, shortly before the publication of the "Zangger" list, India carried out a nuclear test at Pokharan in Rajahstan. This again inflamed Western (and probably Soviet) fears of nuclear proliferation; some saw it as an indication that NPT export controls and the Zangger list were inadequate. Fears of proliferation were also fanned by growing concern that the world was moving towards a "plutonium economy" (thought to be demonstrated by French contracts to sell reprocessing technology to Pakistan and South Korea and German sales of enrichment and reprocessing technology to Brazil) and by the oil crises of 1973-74, which seemed to make nuclear power an attractive alternative source of electricity. These concerns led Kissinger to propose the creation of Nuclear Suppliers' Guidelines (NSG) in order to devise more effective export controls.

With some additions, the NSG incorporated the Zangger trigger list, and introduced the concept of safeguards on "sensitive" nuclear plants built by the use of imported technology (not "safeguards on technology" as some critics maintained). The NSG also brought France - not yet a party to the NPT and not bound by its Article III.2 - into the international system for controlling nuclear exports.

Although the Zangger Committee and the NSG are not international organisations *per se*, they play an important role in deterring proliferation. They have formal international status and informal secretariats, and their work has been noted and occasionally praised by NPT review conferences. In the past, however, both groups - and especially the NSG - attracted the suspicions of many "Third World" spokesmen who saw them as controls designed to bar their access to nuclear energy.

In the 1990s this hostility began to diminish - though not to vanish - as a number of leading developing countries joined the NSG and as the NSG eliminated much of the secrecy that had enveloped its earlier proceedings. Moreover, in 1991 it became clear that controls on nuclear exports had been too lax rather than too restrictive. Saddam Hussein had used several strategies to defeat them: for instance, importing dual-use items not

[311] The Committee was named after its first chairman, Professor Claude Zangger, of Switzerland. In August 1974, ten leading nuclear exporters sent the list to the IAEA and informed it of the steps they were taking to carry out their obligations under Article III.2 of the NPT. See David Fischer and Paul Szasz, *Safeguarding the Atom*, SIPRI, Taylor and Francis, London and Philadelphia, 1985, pp. 101-102.

covered by the Trigger List and using them to make enrichment plant components that were on the Trigger List. To block this loophole the NSG agreed that a large number of dual-use items should henceforth be subject to export licensing (to ensure that they were used for non-nuclear purposes). The NSG also agreed that there should be no nuclear exports to a non-nuclear weapon state unless it had accepted "full-scope" safeguards, that the members of the NSG would refrain from all suspect nuclear exports and that they would meet annually (they had not met since the mid 1970s).

4. The Evolution of IAEA Safeguards

IAEA safeguards now clearly play a central role in deterring nuclear proliferation. This was certainly not the case in earlier decades. The IAEA Statute, which came into force in July 1957, imposed no obligation on any member of the Agency to refrain from making nuclear weapons - in fact France and India, founding members of the IAEA, carried out their first nuclear tests in 1960 and 1974 respectively. Nor did the Statute require any member to accept IAEA safeguards on its own nuclear activities or on nuclear items that it exported. The only exception was that if a state obtained nuclear assistance directly from the IAEA, the imported items and their products (e.g. spent fuel from a reactor supplied by the IAEA) would be subject to IAEA safeguards. But since 1957 only two states, Mexico and Yugoslavia, have turned to the IAEA for supplies of major nuclear plants (power reactors).[312] Moreover, it was relatively easy, at least until the late 1970s, to avoid entirely safeguards, for instance, by developing one's own unsafeguarded nuclear resources or by importing nuclear equipment from a country that did not require safeguards. Thus, in the 1950s, Israel imported a large, unsafeguarded, research reactor and reprocessing technology from France, while Canada supplied India with a similar reactor. This was supplied under Canadian bilateral safeguards, but these fell away when India was able to provide its own fuel in place of the original Canadian uranium.

During the 1960s Western suppliers, led by the US, had begun to require IAEA rather than their own bilateral safeguards (or no safeguards at all!) on their nuclear exports to non-nuclear-weapon states. In 1970, with the entry into force of the NPT, this became an *international* legal requirement. The NPT thus gave great impetus to the spread of IAEA safeguards and enhanced the IAEA's political role.[313] By 1980, all the main industrial non-nuclear-weapon states and most developing countries had accepted IAEA safeguards on all their nuclear activities.

---

[312] In fact, the US firm Westinghouse built the plants. The fact that the IAEA's role as supplier was a legal fiction was embarrassingly demonstrated when in 1978 the US enacted the Nuclear Non-Proliferation Act and forced a protesting Yugoslav government to amend the terms of the relevant agreements so as to bring them into line with the new US law.

[313] In the words of an astute observer, the NPT "*gave the IAEA a tremendous boost...catapulting it from the periphery to the center of the international, political system...*" and transformed it from "*a primarily technical into a frankly political entity*". In doing so, it made the IAEA "*a more inviting political target*". See Lawrence Scheinmann, *The International Atomic Energy Agency and World Nuclear Order*, Resources for the Future, Washington DC, 1987.

The IAEA's safeguards of the mid-1960s were chiefly designed to cover individual plants or shipments of fuel. Under the NPT, however, IAEA safeguards would cover the entire fuel cycle of a non-nuclear weapon state. Accordingly, in 1970-1971, after the NPT came into force, the IAEA found it necessary to draw up a new safeguards system. The chief targets of NPT safeguards were the leading non-nuclear-weapon states, Germany and Japan, still suspect in the eyes of some of their neighbours, particularly since both of them were building plants capable of making nuclear weapon material. Since their industrial competitors in the nuclear-weapon states would not be required to accept safeguards, Germany and Japan also led the non-nuclear-weapon states in insisting that the new system should be less intrusive than its predecessor. Accordingly, the NPT stipulates that the IAEA should make maximum use of (surveillance and other) instruments rather than human inspectors, and that safeguards should apply to nuclear material only and not to plants. Particularly restrictive was the NPT directive, also enshrined in the 1970-1971 safeguards system, requiring that IAEA inspections should normally be confined to a few previously agreed strategic points in the plant concerned.

Nations would not permit international inspectors to search for unreported plants or fuel, and in any case the IAEA would not have enough inspectors for such operations, nor would its member states have provided it with the intelligence needed to make them cost-effective. Hence, the 1970-1971 system focused on the nuclear material *reported to the IAEA by governments*. The system did, however, provide somewhat complex procedures - virtually unused until 1992 - under which the IAEA could send its inspectors anywhere in a non-nuclear-weapon state that had accepted full-scope safeguards.

Despite its limitations the 1970-71 system was able to verify that no diversion of *declared* nuclear material had taken place in any plant in which safeguards were applied. But the disclosure in 1991 of Iraq's extensive nuclear weapon programme showed that the existing system was not capable of detecting even the most extensive clandestine nuclear activities if they were effectively segregated from the nation's declared programme. In the light of these disclosures, the IAEA and its leading members soon introduced some major changes, reaffirming the IAEA's right to send its inspectors anywhere if it appeared that the state was not reporting all its activities, agreeing to provide the IAEA with a much wider range of information about their nuclear activities *and the results of any intelligence activities that indicated a clandestine activity*, and initiating a statement by the Security Council that it would regard any proliferation of weapons of mass destruction as a threat to international peace and security and that its members would take appropriate action if the IAEA reported a violation of a safeguards agreement. Since 1993, these and several other far-reaching reforms have been incorporated in a major revision of IAEA safeguards.[314]

[314] This formally began in 1993 and was expected to be completed in two years, hence its nickname of "Programme 93+2" (it was actually completed in 1997). The official name that the IAEA gives to the modified system is the "Strengthened Safeguards System".

4.1 The Strengthened ("93+2") Safeguards System

The driving motor of the strengthened system is a determination to avoid repeating the Iraqi experience. The strengthened system looks at the activities of the state as a whole rather that at individual plants. In other words, unlike its 1971 predecessor, it is designed to detect diversion by a state rather than by the manager of a nuclear plant. To help the IAEA meet this objective, the state provides the IAEA with information on all its present and future nuclear activities and on past activities, to the extent relevant - all nuclear activities from the mining of nuclear material to the disposal of nuclear waste.

The system also greatly extends the normal access rights of IAEA inspectors; they have virtually unrestricted access to all nuclear and nuclear related plants in the country concerned (and not only to plants that contain nuclear material), access, for instance, to any plant manufacturing nuclear equipment (e.g. components of gas centrifuges) and non-nuclear material (e.g. heavy water), whether or not the plant contains any nuclear material, and to any plant on a nuclear site, whether or not it is "nuclear". The state will also help the IAEA to use new methods of verification such as "environmental sampling" and to make use of the most up-to-date methods of communication including satellites and secure transmission of data recorded by the IAEA's instruments.

The state undertakes to make it easier for an inspector to enter its territory, for instance by waiving visa requirements or granting multiple entry visas. States also agree to simplify and speed up the designation of inspectors.

For its part, the IAEA takes additional measures to protect proprietary information and will seek to reduce routine inspection of run-of-the mill nuclear plants such as light water reactors.

Almost all the leading industrial states, including the EU, Japan, Australia and Canada, have ratified or are in the process of ratifying the protocol embodying the strengthened system. Four of the five nuclear weapon states have agreed to apply relevant aspects of the programme to their civilian nuclear operations.[315] The detailed arrangements for co-operation between the IAEA and EURATOM in applying the strengthened system are now being worked out.

The Director General of the IAEA has set the ambitious target of having the strengthened system accepted by the year 2000 by all states that have comprehensive safeguards agreements. It is particularly important that NPT non-nuclear-weapon states in regions of acute political tension, where the incentive to "get the bomb" is high, should be persuaded to accept the strengthened system, especially those that have already acquired significant nuclear technology such as the two Koreas and Iran.

---

[315] The strengthened system has been accepted by China, France, the US and the UK, and it is expected that Russia will accept in Spring 1999, but ratification by the Duma may take much time. Precisely what nuclear activities (plants and fuel) will in practice be covered by the strengthened system, and how the costs of applying it in the nuclear-weapon states are, is not clear. The IAEA has published the protocol embodying the strengthened system as document INFCIRC/540.

It will, of course, take time to integrate the strengthened system fully into the IAEA's routine safeguards operations. Since it aims at exposing clandestine activities - taking place in an unknown and undefined environment - an absolute assurance that such activities are not taking place will never be attained. But there is little doubt that, if it is vigorously and intelligently implemented, the revised system will greatly enhance the IAEA's ability to detect any clandestine nuclear activities.

Summing up, the strengthened system establishes radically new norms that could be applied in verifying any treaty that limits or bans weapons of mass destruction and that have been shown to be internationally acceptable. The comprehensive information that the strengthened system requires from states, the intrusive inspection that it authorises and its acceptance by all nuclear-weapon and leading non-nuclear-weapon states set precedents that significantly advance the frontiers of verification. Henceforth, it should be more difficult for any state to resist intensive verification on the grounds that it entails an unacceptable infringement of national sovereignty.

5. International Control of Plutonium?

As noted, there was acute concern in the 1970s that the expected widespread production of plutonium would enhance the risks of proliferation. Encouraged (at first) by the US Administration, the IAEA attempted to launch an international plutonium storage system, but no consensus could be reached among those countries that wished to have a system under which plutonium would be released from international storage only if the potential user could make a persuasive economic or technical case for getting it, those that proposed that the only condition for its release should be that the plutonium must remain under IAEA safeguards, and those that wanted prompt and unconditional release of any plutonium separated from their own spent fuel. In the late 1980s, the IAEA Secretariat renewed its consultations on the issue but the states concerned - the main producers and users of separated plutonium - were reluctant to go beyond a voluntary system for international publication of plutonium production and inventories and guidelines for plutonium supply.

Some of the heat has gone out of the issue. In the 1970s and early 1980s many foresaw that plutonium-fuelled breeder reactors would provide the energy of the future. They would step in to meet growing energy needs while reserves of fossil fuels and uranium declined and their price went up. In reality, the demand for electricity rose more slowly than foreseen. Instead of becoming scarce and dear, uranium went into glut. The real cost of fossil fuel fell and the "dash for gas" began. Projections for the growth of nuclear power proved to be quite unrealistic, especially after the Three Mile Island and Chernobyl accidents. France's Superphenix, the first truly large breeder power station in the West, ran into costly and time-consuming technical problems. One by one, all Western European countries abandoned or placed on hold their plans for new breeder reactors. Today, only three countries are pressing ahead with this technology, Japan, the Russian Federation and India. However, the main producers of civilian plutonium,

France, the Russian Federation, the UK and Japan continue to produce the metal and even to expand their production capacity.

The nuclear arms reduction agreements, made possible by the end of the Cold War, led to large surpluses of military stocks of highly enriched uranium (HEU) and weapon-grade plutonium. It was essential that these stocks should be safely stored and disposed of, without risk of proliferation or of their return to military use. Principal responsibility for ensuring safety rests with the states that produced the materials but misuse of weapon-grade uranium and plutonium could pose immense dangers, not only to them, but also to many other nations. In 1997 the US, Russia and the IAEA launched a tripartite study of the matter and the two weapon states agreed to place significant amounts of surplus material under IAEA surveillance.

The problem of surplus highly enriched uranium is simply dealt with by blending it with natural or depleted uranium and releasing the product for use in nuclear power plants. This process does not offer a solution to the plutonium surplus. However, plutonium oxide can be blended with uranium oxide and the product - mixed oxides of plutonium and uranium or MOX - is already being used on a modest scale as a reactor fucl.

The US and Russia have agreed on a two track solution for dealing with the surplus plutonium. It must be made to meet the "spent fuel standard" - i.e. be made as difficult to divert to nuclear explosives as is the highly irradiated spent fuel produced by a nuclear reactor. Two substances meet this criterion - one is MOX fuel and the other is plutonium re-mixed with the nuclear wastes from which it was extracted and vitrified to prevent leaching. On 2 September 1998, Russia and the US announced that each will, stage by stage, irreversibly withdraw about 50 tonnes of plutonium from weapon programmes for conversion into MOX fuel or for agglomeration with high level waste. They would seek to make the process transparent by measures that would include "*appropriate international verification ... and strict standards for physical protection*" and would build the industrial facilities needed for this purpose.[316]

Since the 1980s, when the IAEA last considered the issue of plutonium storage, there have been these two important US and Russian decisions, while the German Government has announced that it will phase out nuclear power. This may lead to the termination of German reprocessing contracts with France and the UK. France has begun to discuss diversifying its purely nuclear electric energy programme. There have been repeated delays in building the large reprocessing plant at Rokkasho-Mura in Japan. The time may thus be ripe to take another look at international plutonium storage and management. In any case, it seems clear that verification of the safe use and disposal of plutonium, military as well as civilian, will play a growing role in the IAEA's programmes in the years ahead.

[316] PPNN Newsbrief, Number 39, 3rd Quarter 1997, p. 3, and the Russian presidential press service web site, 2 September 1998, quoted by the BBC Monitoring Summary of World Broadcasts.

6. The International Bodies Administering the CWC, BWC and CTBT

Only two international treaties, the Chemical Weapons Convention (CWC) and the NPT are, as yet, buttressed by fully functioning organisations to verify that their parties are complying with the crucial provisions of the treaties. Negotiations are proceeding on the creation of a body to verify the Biological Weapons Convention (BWC). The entry into force of the Comprehensive Test Ban Treaty (CTBT) and the formal creation of a CTBT Organisation (CTBTO) are currently blocked by the refusal of some of the CTBT's most crucial potential parties, e.g. India and North Korea, to ratify the treaty except, in India's case, within the framework of total nuclear disarmament.

The CWC and the BWC aim permanently to ban all chemical and biological weapons. In this sense, they are akin to the Intermediate-range Nuclear Forces (INF) Treaty, and to the Strategic Arms Reduction Treaties (START), which ban all weapons within certain categories or ranges. They are even more akin to an eventual global nuclear weapons convention that would aim at banning all nuclear weapons. They thus differ from the NPT, which not only seeks to maintain and promote nuclear technology but also tacitly permits five states to maintain - and expand - their nuclear arsenals until an undefined future date.

These differences of aim, as well as technology, lead to significant differences in the verification procedures and techniques which the treaties use. The first aim of the CWC (as well as for the INF, the START treaties and a global nuclear weapons convention) is to verify the complete elimination of all existing weapons in the categories they target. At the same time, the CWC and the other listed treaties must verify that no new weapons in the targeted categories are produced. The single aim of the NPT (in relation to nuclear disarmament) is to verify that nuclear technology is used only for civilian purposes, firstly by the non-nuclear weapon states that join the treaty and, in due course, also by the nuclear-weapon states.

One consequence is that once the first aim of the CWC, INF and START treaties and of a global ban on nuclear weapons is achieved, i.e. once the targeted weapons have been eliminated, it should be possible to reduce substantially the resources required for verification. In the case of the NPT, however, no such reduction can be foreseen even when the final aim of its Article VI is achieved, since it will still be necessary to verify that there is no diversion of nuclear material from civilian (or permitted military) uses.[317]

Verification under the NPT also relies more than under the other treaties and organisations on *routine* accounting and inspection (and on the "complementary access" foreseen under the strengthened system). Such inspection is carried out by the technical secretariat of the IAEA on its own responsibility and initiative rather than (as in the case of the CWC and CTBT) chiefly by "challenge" or special inspections carried out at the behest of another state or of the (inevitably) highly politicised governing body of the

---

[317] Article VI: "...a treaty on general and complete disarmament under strict and effective international control".

organisation concerned.[318] In this sense, the IAEA's verification of non-diversion resembles the work of a government agency overseeing industrial safety rather than that of a military or police organisation.

Nonetheless, the verification tasks of the agencies have some important features in common. This applies particularly to the tasks of verifying the NPT and a CTBT. In both cases, detection of certain radio-nuclides plays a role. Such detection may be crucial in the case of verifying the absence of nuclear diversion, it is secondary to verifying the absence of nuclear tests. More to the point, a nuclear test would not only be a violation of the CTBT, if carried out by a non-nuclear-weapon state, it would also be the most unequivocal demonstration of nuclear proliferation and of a violation of the NPT.

7. A Cut-off Convention

Short of agreement on the elimination of weapons of mass destruction, one of the most significant measures leading to total elimination would be an internationally verified convention putting an end to the production of fissile material for nuclear weapons. For reasons of economy and for the most effective use of manpower, the IAEA would not apply safeguards to power and research reactors in the nuclear-weapon states (for instance to the 110 light water power reactors in the US and the 33 in Russia) but would concentrate its safeguards on the "sensitive" nuclear facilities in those states - in other words, on the (relatively few) reprocessing and enrichment plants that were kept in operation for the production of civilian plutonium or low enriched uranium.[319]

If, however, such a regime is confined to the nuclear-weapon states, it seems likely that the non-nuclear weapon states will press for the same treatment. A possible (partial) solution would be to continue existing EURATOM safeguards on all civilian nuclear plants in the EU, while the IAEA (together with EURATOM) would only safeguard their "sensitive" plants. A similar two-tiered regime might apply in other cases where there are well-established and tested regional nuclear energy systems (Argentina/Brazil and their "Brazilian-Argentine Agency for Accounting and Control of Nuclear Materials", ABACC) or effective national systems of accounting and control as in Australia, Canada and Japan.

Presumably, a cut-off convention would require verification of the dismantling of remaining military reprocessing or enrichment plants, or their conversion to civilian use,

---

[318] The organisation for the prevention of chemical weapons also carries out routine inspections, for instance to verify that industrial chemical plants are not being used to produce chemical warfare agents.

[319] Over the years, the IAEA has acquired considerable experience in safeguarding such plants, for instance reprocessing plants in Germany (WAK), in Japan (Tokai), in India from time to time (PREFRE), and some experience in the UK (Sellafield), and in France (La Hague), and gas centrifuge enrichment plants in the UK (Capenhurst), in the Netherlands (Almelo), in Germany (Gronau) and in Japan (e.g. Rokkasho), and some experience in safeguarding other types of enrichment plants in South Africa, Brazil and Argentina. It seems that the recognised nuclear weapon states have ample stocks to meet the needs of their nuclear-engined submarines and other naval vessels. Since no text of a cut-off convention has yet been tabled this is still uncertain; equally uncertain is the question whether a cut-off convention would require the application of safeguards to other *existing* stocks of highly enriched uranium and military plutonium.

and the application of IAEA safeguards to any civilian plants whose throughput was not already under safeguards. In all cases, the application of the strengthened system together with traditional IAEA safeguards would be the rational choice.[320]

8. The Role of the Security Council

All the verification agencies rely on the Security Council as the final tribunal to which they would turn in the event that the treaties they administer were violated. This reflects the fact that the Security Council is the only United Nations body that has the express legal authority to enforce compliance, even by military measures,[321] with its decisions and with international treaties. During the Cold War the Council was usually prevented from doing so by the vetoes of the permanent members. In 1990, after Iraq invaded Kuwait, the Council used its authority to call effectively for the use of armed force for the first time since the early 1950s.[322] After 1992, when it was determined that North Korea had violated its IAEA safeguards agreement, the Security Council invariably voiced support of the IAEA's attempts to implement the agreement, but fear that one of its permanent members, China, would use its veto apparently dissuaded the Council from imposing stronger sanctions.

Unless there is a fundamental change in the way the Council operates, it is clear that its effectiveness will depend on achieving effective consensus amongst the permanent members. This may not be a very satisfactory conclusion but it has been the reality of international politics since 1945.

A state that has already violated one of the treaties dealing with weapons of mass destruction may be more likely to violate other treaties in this category - a case in point is Iraq. The verifying agencies thus have a strong interest in ensuring compliance with all verification treaties, in ensuring that any non-compliance is promptly remedied, and that continued defiance of a treaty provokes effective sanctions. The agencies also have an interest in avoiding conflicting interpretations of such treaties.[323] Given the different political composition of the agencies and of the regional bodies concerned, such conflicts,

---

[320] It is understandable that some proponents of nuclear disarmament would wish a cut-off convention to cover all existing stocks, as well as future production of the material. However, if "existing stocks" were taken to include fissile material already fabricated into the explosive component of nuclear warheads, this would imply total nuclear disarmament, which, however desirable, cannot be achieved as an indirect by-product of another nuclear arms control measure and would be vetoed by the NPT nuclear-weapon states. If existing stocks meant less than this, the precise scope of the convention might be uncertain and its definitions might require lengthy negotiations, even if such a convention were acceptable to the *de jure* and *de facto* nuclear weapon states. The course of practical wisdom is probably to confine the convention to future production of fissile material and to leave the door open for future extension of its scope.

[321] UN Charter, Chapter VII, Article 42.

[322] Generally speaking the Council demonstrated its willingness and ability to enforce its decisions in the case of Iraq, though its cohesion has weakened with the passage of time. The record of the Security Council in the Balkans is more mixed and it is still being put to a test, but in this case the issue does not relate to weapons of mass destruction.

[323] For instance, one of the verification agencies determines that a ban on nuclear tests is about to be violated, while another agency is not able to conclude that there has been a diversion of nuclear material.

although unlikely, are not impossible. This points to the need for close co-operation between the agencies on techniques and practices of verification and on matters of policy.

9. Some Conclusions

9.1 Nuclear threats

Without agreement on the total prohibition of nuclear weapons, the NPT, CTBT as well as the proposed cut-off convention are painfully vulnerable to the (unjust) charge that they are designed to do no more than entrench the privileged status of the recognised nuclear-weapon states. Such a far-reaching prohibition is, however, not yet in reach and there is no indication that it will be in the foreseeable future. The nations of the world must therefore make best use of the international instruments they already have and seek to complement them with those that are now possible. The obvious top priorities are to bring the Comprehensive Test Ban Treaty into force without delay and to negotiate and bring into force a cut-off of the future production of nuclear material for nuclear weapons.

9.2 The IAEA

The Agency is likely to focus on:

- bringing the strengthened safeguards system into full operation. The IAEA's Director General aims to have strengthened safeguards accepted by all NPT non-nuclear-weapon states by the year 2000;
- making greater use of regional and national safeguards systems and reducing routine safeguards at run-of-the mill plants (e.g. light water reactors);
- (hopefully) promoting and verifying a cut-off convention;
- if such a convention is concluded, applying a two-tiered safeguards regime in nuclear-weapon states and possibly in states covered by effective regional or national safeguards systems;
- verifying that civilian fissile material, as well as that released from military stocks, is safely stored and irreversibly kept in civilian use, or is disposed of as nuclear waste. IAEA verification of the actual dismantling of nuclear warheads will depend largely on developing effective means of preventing the diffusion of information about warhead design;
- re-examining the issues involved in international plutonium storage and management. In any case, verification of the safe use and disposal of plutonium, military as well as civilian, will play a growing role in the IAEA's programmes.

If a cut-off convention comes within reach, or if the EU expands further, it will become even more desirable to find a way of ending unnecessary duplication between the verification activities of EURATOM and the IAEA.

9.3 Chemical and biological threats and the need for a complete elimination of nuclear weapons

A for chemical and biological threats, the most urgent need is to create effective machinery for verifying the biological weapons convention. Many have drawn attention, however, to the absurdity of imposing a total ban on certain weapons of mass destruction (chemical and biological), while a handful of states can legally retain and manufacture the most destructive of all weapons in this category, the nuclear weapon.

9.4 Co-operation between the verification agencies and the Security Council

Formal status should be given to the President of the Council's declaration of 31 January 1992 that the Council would regard any proliferation of weapons of mass destruction as a threat to international peace and security. Should there be more formal arrangements for consultation and exchange of experience between the secretariats of the verification agencies - IAEA, the Organisation for the Prevention of Chemical Weapons (OPCW) and the interim secretariat of the CTBTO - including the secretariats of the regional bodies? And arrangements for regular briefing of the Security Council? What can be done to encourage a more robust and predictable reaction by the Council to breaches of arms control agreements?

Concluding Impressions

by
C.R. Hill, A. Mechelynck, G. Ripka and B.C.C. van der Zwaan

This book has been prepared within the framework of a workshop organised by the *Pugwash Conferences for Science and World Affairs*. It is however the clear policy of Pugwash that any documented outcome of one of its workshops is the sole responsibility of the participants. In the course of this particular workshop meeting the various authors and other participants specifically decided not to draw up a formal set of overall conclusions. There seem, nonetheless, to be a number of threads that can usefully be drawn together from the various chapters in the book, and from the critical discussions that were held on them. The following attempt to do this is thus purely the responsibility of the editors and does not carry any endorsement either by Pugwash or by the other workshop participants.

Global Energy Management

Major problems in energy management will occur in the coming century, and these now need to be addressed urgently. The combination of projected population increase and raising of GDP expectations, particularly in developing countries, suggests a factor of about 5 increase in global energy demand by 2100. Improved energy efficiencies can and must be achieved but are unlikely to diminish substantially the above prediction. Satisfaction of this increased demand for energy will call for increased use of some or all of: fossil carbon fuel, nuclear energy, and "renewables".

Fossil fuel at present supplies some 80 % of world energy needs. It generates substantial CO_2, particulate and other environmental pollution, particularly in countries where the "poverty trap" mitigates against clean and efficient technique. Fossil carbon is a finite resource: oil and gas will become scarce in the coming century; coal will last somewhat longer. The evidence for global warming, attributable substantially to man-made CO_2 emissions, is compelling. The exact consequences of such global warming are unknown. They may comprise a combination of detriment and some eventual benefit. There must, however, be a serious likelihood of substantial human suffering, environmental damage, and sources of conflict.

Renewables that may make a substantial, non-greenhouse, contribution include: hydro, wind power and solar energy (both through direct heating and *via* photo-electric conversion). Reliable assessment of the full potential of most of these renewables, or of the use of "decarbonised" fossil fuels, will require substantially increased investment in research, development and pilot trials. At present, however, it is very doubtful whether, in any combination, they have the potential to provide the bulk of projected 2100 world energy requirements. Thus, in their forward planning, it is unrealistic to expect governments to place total reliance on such resources.

Nuclear fusion could, eventually, prove to be an economically and environmentally acceptable means for providing a substantial, or even predominant, fraction of world energy needs. It may, however, be at least several more decades before we know whether its implementation is technically and economically practicable. In 1995, nuclear fission provided 18 % of the world's electrical power (7.3 % of total energy). The geo-political variation of that provision is considerable: from some 80 % in France, through 15-50 % in the majority of other industrialised countries, to zero in many of the poorer or smaller developing countries.

Objective comparison of the costs of nuclear- and fossil-generated electricity is difficult, particularly when taking into account the real environmental costs (including decommissioning) of both technologies. Undoubtedly, much of the pseudo-commercial viability of nuclear electricity in its early days was artificially supported by being able to write off much of the development costs against weapons budgets. Against this, however, there are now several examples of industrial economies (again, in particular France) that appear to operate very successfully whilst relying heavily on nuclear power.

Commercial interests have played a major part in the development of nuclear energy, although not always for the best. In the UK, for example, commercial forces have driven the momentum for reprocessing far beyond its overall economic (let alone security) justification. Conversely, enlightened public-sector leadership of the reactor programme in France appears to have resulted in economies of scale that have not been matched in countries where, as in the US, the open commercial market has had free rein.

There may be a strong case for well-controlled commercial and/or World Bank investment, in developing countries, in projects designed to enhance energy efficiency. A characteristic of many developing countries is the relative fragility of their electrical power distribution systems. This, often taken together with a shortage of suitable manpower and access to affordable capital, can present major problems in the introduction of nuclear electricity. Nonetheless, a number of such countries (particularly those that are oil-poor and thus wish to reduce reliance on external fuel supplies, such as China, India and Pakistan) have already established substantial, and largely home-grown nuclear power programmes.

Thus, whilst major questions exist over nuclear energy, it has become well established world-wide, in some cases apparently to the point of being indispensable, as one component of energy supply, at least during the next few decades. There are powerful arguments for increased development and trials of renewables, but unless and until such sources have been shown to be capable adequately to provide for world energy needs in the coming century, it will not be possible to rule out the necessity to continue and expand the nuclear contribution.

Health and the Environment

The normal process of the generation of electricity from nuclear fission (in common with that from fossil fuel burning) results in the release of pollutants into the environment. The nature of nuclear (radioactive) pollutants is that they are expected to add to the existing natural incidence of cancer induction and genetic change in humans and other species. The extent of release of such pollution (both absolutely and per unit of generated electricity) has been reduced substantially in the past 10 years by improvements in technology, but may never be eliminated altogether.

In terms of biologically damaging doses to the typical human being, current nuclear electricity generation is expected (on the basis of United Nations' figures) to add some 0.01 % to the effect of existing, naturally occurring, radiation. At a conservative estimate, this translates to an expectation of some 100 additional cancer deaths annually, world-wide, over and above the approximately two million that are believed to be attributable to natural radiation.

In many countries it has not been possible to establish a settled, long term policy for managing the radioactive waste materials that result from reactor operations. Current practice, therefore, is to place materials in essentially temporary, shallow storage. Escapes of such material have occurred, either through bad management or resulting from unforeseen accidents. Among long-term options for disposal of high level wastes, geologically deep storage is generally favoured. There is increasingly good evidence that this could be implemented without consequences that would significantly add at any stage to human exposure to damaging natural radiation. On this topic, however, there exists a strong and understandable "Not In My Back Yard" syndrome, which has had effective political outcome in blocking implementation.

Nuclear power operations may also pollute the environment through major accidents, as was vividly illustrated by the Chernobyl accident, which was much the most damaging reactor accident to have ever occurred. An immediate human consequence of the Chernobyl accident was the 31 acute radiation deaths of operating and safety personnel in the month following the accident. In addition, the WHO and OECD currently predict, as a consequence of the accident, an 0.01 % lifetime increase in cancer incidence averaged over the European population (but predominantly in individuals who were immediately down-wind at the time of the accident), implying an excess premature mortality of the order of 7000. Furthermore, there might be a similar number of adverse genetic changes affecting future generations. Additionally, there were major non-radiation effects: psychological, social and economic trauma, throughout a large population.

The main causes of the Chernobyl accident appear to have been serious human failure and unsafe design. The unit involved, a RBMK graphite moderated, water cooled reactor, was of a design that originated in the Soviet Union and is recognised as lacking safety features that are now standard elsewhere, where water-cooled, water moderated designs are now almost universal. It is noteworthy that there is now some 9000 reactor-years experience of running such water reactors, in which there has only been one substantial accident: that at Three Mile Island, as a consequence of which there was no recorded or predictable human injury. Reactor safety remains a major cause for concern, however, and various proposals have been made in the direction of decreasing (albeit at some cost) risks and consequences of accidents below those inherent in current practice.

Whatever the validity of the above statements, and in whatever perspective they are viewed, there is a strong perception, among substantial groups in many countries, that the health and environmental risks posed by nuclear power are absolutely unacceptable. This perception is not universal, and seems to be largely characteristic of relatively prosperous, industrialised societies. In any case, from the point of view of optimal global energy management, it will be important to assess whether this perception is based on objective evidence and logical analysis.

Nuclear Weapons Proliferation

For many people, the gravest concern over nuclear energy arises from its incidental generation of weapons-usable fissile material. In many situations, production of highly enriched uranium (HEU) may be the most practical route for illicitly acquiring weapons-usable fissile material. However, much current concern is focused on production of ^{239}Pu in uranium burning reactors. Furthermore, thorium burning reactors, that would produce fissile ^{233}U, are under design in some countries.

For ^{239}Pu there are conflicting arguments as to whether security is maximised, on the one hand, by storing it as a constituent of spent fuel or, on the other, by extracting it from waste (reprocessing), followed by burning it as a mixed Pu/U oxides fuel ("MOX"). The reprocessed ^{239}Pu from normal power reactor operation, although sub-optimal for weapons use, is nonetheless considered to be weapons-usable. World-wide, there are stocks of some 525 tonnes of plutonium (as metal or oxide) that have arisen either from power reactors or weapons decommissioning. This is being added to by continuing reprocessing of spent fuel. There are thus serious concerns for maintaining security of these stocks and for finding means (e.g. by burning as MOX or by mixing with stored spent fuel) for reducing and ultimately eliminating them.

Amongst technologies that could have the potential to reduce plutonium stocks is the fast-neutron reactor, that can be operated as an energy-generating plutonium burner. Developments here, however, have largely been discontinued, primarily on grounds of forecast economic viability. Another such technology that is being explored is the hybrid "proton accelerator driven reactor". This would operate sub-critically (thus supposedly being inherently rather safe) and would function both as a power generator and incinerator of plutonium and other actinides. Its feasibility and economics are yet to be demonstrated.

A substantial system of international and regional safeguard mechanisms has been established in an attempt to contain the problem of misuse of fissile materials. This appears to be largely effective, but it still contains two gaps: the five "original" nuclear weapon countries, that are only partly covered; and several nominally non-nuclear weapon countries, some of whom may be conducting undeclared activities that are in breach of the NPT. On the grounds that the nuclear weapon powers already have more plutonium and HEU than they know what to do with, it is particularly to these latter undeclared activities that further safeguarding needs to be directed.

Appendix: Technical Notes and Units

by
André Mechelynck*

Introduction

Several different units are used in this work for the same physical quantities, such as power and energy. At many instances, and where feasible, the SI equivalents, i.e. the "International System" units, have been mentioned in the text (between brackets).[324] In addition, the editors have thought it useful to provide the reader with the following data and conversion tables. The numbers given are limited to a maximum of four significant figures.

1. Prefixes

Prefix	Symbol	Value	
micro	μ	0.000 001	10^{-6}
milli	m	0.001	10^{-3}
kilo	k	1 000	10^{3}
mega	M	1 000 000	10^{6}
giga	G	1 000 000 000	10^{9}
tera	T	1 000 000 000 000	10^{12}
peta	P	1 000 000 000 000 000	10^{15}
exa	E	1 000 000 000 000 000 000	10^{18}

NB: "billion" may, according to local use, mean either 10^9 (1 000 000 000) or 10^{12} (1 000 000 000 000). In the current work, it should always be construed as 10^9.

2. Time

The basic unit is the *second* (s).
1 *hour* (h) corresponds to 3.6 kiloseconds (ks).
1 *day* (d) corresponds to 86.4 ks.
1 *year* (y) corresponds to 365.25 d or 31.56 Ms.

---

* André Mechelynck is a Consultant on Energy Matters and former Engineering Advisor.

324 Some of the basic units of the SI system are the meter (m), the kilogram (kg), the second (s) and the Kelvin (K).

3. Mass

The basic unit is the *kilogram* (kg). The most used derived unit is the *ton* [325], (t or Mg, 1000 kg). It should not be confused with either the *Anglo-Saxon ton* (2240 lbs, or about 1016 kg) or the *short ton* (2000 lbs, or about 907 kg). In this work, *ton* always means 1000 kg.

4. Temperature

Temperatures are normally stated, for convenience, in *Celsius* (C), although the International System unit is the *Kelvin* (K). The classical Anglo-Saxon scales (*Fahrenheit* and *Rankin*) have not been used in the present work.

For the record:
1 K = 1 C = 1.8 F = 1.8 R

Origins of scales:

0 K corresponds to:	0.00 R	(absolute zero)
	- 273.16 C	
	- 459.69 F	
0 C corresponds to:	273.16 K	(freezing point of water)
	32 F	
	491.69 R	

5. Pressure

The SI unit is the *pascal* (Pa, 1 N/m^2, *newton* per square meter).[326]
1 *bar* equals 10^5 pascal.[327]

---

[325] Also spelled as *tonne*.

[326] The *newton* (N) itself is the force that communicates to a mass of one kilogram (kg) an acceleration of one meter per second squared ($m.s^{-2}$). The standard acceleration of gravity is equal to 9.80665 $m.s^{-2}$, thus 1 *kilogram.force* is equal to 9.80665 N.

[327] This unit *bar* is convenient, since its value is close to that of the *standard atmosphere* (101 325 Pa) and of the old *kilogram.force per square centimeter* (kgf/cm^2, 98066.5 Pa).

6. Energy[328]

The basic unit is the *joule* (J), the work executed by a force of one newton (N), displacing its point of application by one meter (m).

Commonly used units mentioned in this volume are:

- the *watt.hour* (Wh, 3.6 kJ) and its multiples: kWh, MWh, GWh, etc.
- the *watt.day* (Wd, 86.4 kJ) and its multiples.
- the *watt.year* (Wy, 31.56 MJ) and its multiples.
- the *kilocalorie* (kcal), equal to 4.186 kJ.
- the *electron-volt* (eV), equal to (approximately) $160.2\ 10^{-21}$ J.
- the *British thermal unit* (Btu)[329], equal to 0.252 kcal or 1.055 kJ.
- the *tonne of bituminous coal unit* or *tonne of coal equivalent* (TBCU, TCE), which is roughly equivalent to 31.56 GJ (1 kWy).
- the *tonne of oil equivalent* (TOE), which is roughly equivalent to 41.86 GJ.
- the *m³ of wood* (stere, st), equivalent to about 6.5 GJ.

The combustion of 1 kg of pure carbon releases about 37 MJ.
The fission of 1 kg of ${}^{235}_{92}U$ releases about 60 TJ.
The fusion of 1 kg of fusion fuel (${}^{2}_{1}H$, ${}^{3}_{1}H$) releases about 380 TJ.
One *barrel of petroleum* (actually a measure of volume) is equal to 42 gallons (US), equivalent to 0.159 m³, and will yield about 6.6 GJ (± 10 %).

The above figures yield the following conversion table:

	J	**kWh**	**Wy**
1 J	1	$277.8\ 10^{-9}$	$31.69\ 10^{-9}$
1 kWh	$3.6\ 10^{6}$	1	$114\ 10^{-3}$
1 Wy	$31.56\ 10^{6}$	8.77	1
1 kcal	4 186	$1.16\ 10^{-3}$	$133\ 10^{-6}$
1 Btu	1 055	$293.1\ 10^{-6}$	$33.4\ 10^{-6}$
1 m³ wood	$6.5\ 10^{9}$	1 805	206
1 TBCU	$31.56\ 10^{9}$	8 767	10^{3}
1 TOE	$41.86\ 10^{9}$	$11.63\ 10^{3}$	$1.326\ 10^{3}$
1 kg ${}^{235}_{92}U$	$60\ 10^{12}$	$16.7\ 10^{6}$	$1.90\ 10^{6}$

[328] A distinction must be made between heat and other (noble) forms of energy (kinetic and potential energy, electricity, mechanical energy, etc.). Where necessary, we specify W(e) or W(th) to indicate, respectively, electrical or thermal energy. In a modern power plant, it takes about 3.3 J(th) of heat to produce 1 J(e) of electricity.

[329] Defined as the amount of heat needed to raise the temperature of 1 lb of water by 1 F at or around 39.1 F.

7. Power

The basic unit is the *watt* (W), equal to one joule per second.
1 kW corresponds to 860 kcal/h or 20 640 kcal/d.
1 TWh/y corresponds to 114 MW.
1 EJ/y corresponds to 31.69 GW.
1 TBCU/y (1 TCE/y) corresponds to 1 kW.

1 m² of photocells absorbs about 250 W, during daylight. Power production yields in the neighbourhood of 10 %, on a yearly average basis, should be expected, i.e. 25 W/m².

	kW	**kcal/h**	**TWh/y**	**EJ/y**
1 kW	1	860	$8.77\ 10^{-6}$	$31.56\ 10^{-9}$
1 kcal/h	$1.1627\ 10^{-3}$	1	$8.94\ 10^{-9}$	$36.6\ 10^{-12}$
1 TWh/y	114 000	$111.8\ 10^{6}$	1	$3.60\ 10^{-3}$
1 EJ/y	$31.69\ 10^{6}$	$27.3\ 10^{9}$	278	1

8. Carbon and Carbon Dioxide

a) Quantities

Basic unit: the kilogram of carbon (kgC) and its multiples: MgC (metric ton, tC), GgC, TgC, etc.
One kg of carbon corresponds to 3.67 kg of CO_2.
One kilomole represents 12 kgC and 44 kg of CO_2.
One part per million (ppm), *in volume*, of CO_2 in the atmosphere corresponds to a little less than 2 mg (milligram) per m³ of air, and to roughly 1.85 GtC in the whole atmosphere.[330]

b) Flows

1 t/y of carbon corresponds to 3.6 g/s of carbon.
The current annual flow of carbon to the atmosphere is roughly 8.4 Gt/y of carbon, from the following sources:[331]

[330] Based on about $5.9\ 10^{18}$ Nm³ (standard cubic meter) for the concerned atmospheric volume.
[331] For more details, please refer to Chapter 2, by Fetter.

Oil	2.7 GtC/y	32 %
Coal	2.3	24 %
Natural gas	1.36	16 %
Gas flaring	0.05	0.6 %
Cement manufacture	0.15	1.8 %
Volcanoes, etc.	0.10	1.2 %
Land clearing	1.50	18 %
Fuel-wood, etc.	0.25	3.0 %

c) World inventory[332]

Atmosphere	760 GtC	(CO_2, carbon compounds)
Biota	700 GtC	(living matter)
Oceans (dissolved)	38 000 GtC	(CO_2 and carbonates in solution)
Soil (organic)	1 400 GtC	(coal, oil, gas, etc.)
Soil (inorganic)	30 000 000 GtC	(limestone rocks)

9. Energy Sources and Use

Source	Production 1996 (1)	Recoverable Reserves (1)	Reserves (years) (a)	Consumption 1996 (1)
Coal and Peat	4607 Mt	1031 Gt	224	2257 MTOE
Gas	2009 MTOE	129 GTOE	64	1972 MTOE
Oil	3362 MTOE	141 GTOE	42	3312 MTOE
Uranium	36 kt (2)	3.4 Mt (b)	100	621 MTOE

(1) BP statistical Review 1997.
(2) NEA/OECD and IAEA, *Red Book*, 1997.
(a) Reserves/Production 1996.
(b) At an extraction price < \$130 / kgU. In chapter 1, a value of 4.4 Mt is quoted, at the same extraction price. Different sources in the literature give values varying around 3 to 4 Mt.

10. Economies of Energy

As an example of the possibility of energy savings in households, the following table shows the evolution of the average energy use of household appliances from 1980 to 1995.[333]

[332] John F. Holdren, private communication, August 1998.
[333] Electrabel, *Energie*, December 1998, p. 3.

Household Appliance	Average Energy Use (W, normal use)	
	1980	**1995**
Refrigerator	40	27
Refrigerator with freezer compartment	53	37
Deep freezer (300 dm³)	81	56
Deep freezer (420 dm³)	105	70
Washing machine (5 kg)	125	92
Drier (5 kg)	158	133
Dishwasher	104	71

11. Radioactivity[334]

becquerel (Bq) : Quantity of radioactive element undergoing one atomic disintegration per second.
gray (Gy) : 1 J of radiation energy absorbed by 1 kg of tissue.
rad : 0.01 Gy.
sievert (Sv) : Amount of radiation equivalent in biological "effectiveness" to 1 Gy of gamma rays.
rem : 0.01 Sv.

At sea level in the temperate zone, the mean exposure to an average member of the US population is about 0.27 mSv/y from cosmic radiation, 0.28 mSv/y from natural radioactivity in soil or rock (depending strongly on the kind of soil, the region, etc.), and about 0.36 mSv/y from radioactive elements (Rn, Ra, K, Ca) deposited internally, for a total of 0.91 mSv/y. To this must be added about 1.06 mSv/y of man-made origin (including mainly medical diagnostic and radio-therapy contributions, in addition to very small amounts from nuclear power plant emissions), resulting in a total of about 2 mSv/y.[335] The figure for the world average of the total of natural background and man-made radiation contributions (UNSCEAR 1988), as quoted in chapter 1, is slightly higher: 2.7 to 3.4 mSv/y.

[334] For the sake of completeness, we mention here some older units: the *curie* (Ci), equal to 37 10^9 Bq (approximately 1 g of Ra); the *rutherford*, equal to 10^6 Bq; the *röntgen* (R), equivalent to 9.3 10^{-3} Gy.

[335] Encyclopædia Britannica, 15th ed., 1991, *Radioactivity*, Vol. 26, p. 488. Man-made radiation comes essentially from medical and dental diagnostic procedures. Atmospheric nuclear bomb tests in the past have resulted in a small man-made radiation contribution. As for industrial emissions, nuclear power plants in normal operation contribute only a very small amount to the total man-made radiation. In fact, radioactive emissions from nuclear power plant are normally of the same order as those produced by an equivalent coal-burning plant, since coal contains various radioactive elements. The Three Mile Island 1979 accident released 0.8 mSv within a 16 km radius, and 0.015 mSv within a 80 km radius (Enc. Brit., *op. cit.*, *Poisons and Poisoning*, Vol. 25, p. 925).

12. Element Symbols

In this work, elements symbols are always shown as $^{Z}_{N}A$, where:

A is the element symbol (H for hydrogen, and so forth),
N is the atomic number (number of protons in the nucleus),
Z is the mass number (total number of protons and neutrons).

For instance, the fissile isotopes of uranium are presented as $^{235}_{92}U$ and $^{233}_{92}U$. The index N is sometimes, for convenience, omitted.

List of Abbreviations

ABACC	Brazilian-Argentine Agency for Accounting and Control of Nuclear Materials
ABM	Anti Ballistic Missiles
ADS	Accelerator-Driven Systems
AEC	American Atomic Energy Commission
AERB	Atomic Energy Regulatory Board
AGR	Advanced Gas-cooled Reactor
AHWR	Advanced Heavy Water thermal Reactor
AIMBY	All In My Back Yard
ALARA	As Low As Reasonably Achievable
ALARP	As Low As Reasonably Practical
ANS	American Nuclear Society
APS	American Physical Society
ASIATOM	Asian Atomic Energy Agency
ASTM	American Society for Testing Materials
ATW	Accelerator Transmutation of Waste
AVM	*Atelier de Vitrification de Marcoule*
BARC	Bhabha Atomic Research Centre
BNFL	British Nuclear Fuel Limited
BRIT	Board for Research in Isotope Technology
BWC	Biological Weapons Convention
BWR	Boiling Water Reactor
CANDU	Canadian Deuterium Uranium reactor
CAPRA	*Consommation Accrue de Plutonium dans les réacteurs à neutrons Rapides*
CEA	*Commissariat à l'Energie Atomique*
CEGB	Central Electricity Generating Board
CERN	European Laboratory for Particle Physics
CFCa	*Centre de Fabrication de Cadarache*
CIS	Commonwealth of Independent States
CISAC	Committee on International Security and Arms Control
CNRS	*Centre National de la Recherche Scientifique*
COGEMA	*Compagnie des Matières Nucléaires*
CSIS	Center for Strategic and International Studies

CTBT	Comprehensive Test Ban Treaty
CTBTO	CTBT Organisation
CWC	Chemical Weapons Convention
DOE	United States Department of Energy
ECN	Netherlands Energy Research Foundation
ELECTRABEL	Belgium's largest electricity company
EU	European Union
EURATOM	European Atomic Energy Agency
FBR	Fast Breeder Reactor
FBTR	Fast Breeder Test Reactor
FCCC	Framework Convention on Climate Change
FSU	Former Soviet Union
FY	Fiscal Year
GDP	Gross Domestic Product
GNP	Gross National Product
GWP	Gross World Product
HEU	Highly Enriched Uranium
HLLW	High Level Liquid Waste
HLW	High Level Waste
HTGR	High Temperature Gas Reactor
IAEA	International Atomic Energy Agency
ICRP	International Commission on Radiological Protection
IFR	Integral Fast Reactor
IIASA	International Institute for Applied Systems Analysis
ILW	Intermediate Level Waste
INEA	International Nuclear Energy Academy
INF	Intermediate-range Nuclear Forces treaty
INSAG	International Nuclear Safety Advisory Group
IPCC	Intergovernmental Panel on Climate Change
ITER	International Thermonuclear Experimental Reactor
JAERI	Japan Atomic Energy Research Institute
kgHM	kilograms of Heavy Metal
LBE	Lead-Bismuth Eutectic
LEU	Low Enriched Uranium
LINAC	Linear Accelerator
LLW	Low Level Waste
LWR	Light Water Reactor
MA	Minor Actinides
MDF	MOX Demonstration Facility at Sellafield
MHTGC	Modular High Temperature Gas Cooled Reactor

MINATOM	Russian Ministry of Atomic Energy
MIX	MIxed oXide, plutonium + slightly enriched uranium, $(U,Pu)O_2$
MOX	Mixed OXide, plutonium + depleted or natural uranium, $(U,Pu)O_2$
MTOE	Million Tons of oil Equivalent
MWd/ton	Megawatt-days per ton
NAS	US National Academy of Sciences
NGO	Non-Governmental Organisation
Nirex	Nuclear Industry Radioactive Waste Executive
NNWS	Non Nuclear Weapons States
NPT	Non-Proliferation Treaty
NRC	Nuclear Regulatory Commission
NRDC	US Natural Resources Defense Council
NSG	Nuclear Suppliers' Guidelines
NWS	Nuclear Weapons States
OECD	Organisation for Economic Co-operation and Development
OECD / NEA	OECD / Nuclear Energy Agency
OPCW	Organisation for the Prevention of Chemical Weapons
PCPA	Post-Closure Performance Assessment
PHWR	Pressurised Heavy Water Reactor
PIUS	Process Inherent Ultimate Safety
P-T	Partitioning-Transmutation
PWR	Pressurized Water Reactor
RAWMAC	Radioactive Waste Management Advisory Committee
RBMK	*Reaktor Bolshoi Moshnost Kanaly*: "high power channel reactor"
RCF	Rock Characterisation Facility
R&D	Research and Development
R-Pu	Reactor-grade Plutonium
R&R	Reprocessing and Recycling
SDI	Strategic Defense Initiative
SIR	Safe Integral Reactor
SMP	Sellafield MOX Plant
SRP	Savannah River Plant
START I, II, ...	Strategic Arms Reduction Treaties
TCE	Ton of Coal Equivalent
T&D	Transmission and Distribution
THORP	Thermal Oxide Reprocessing Plant
TIFR	Tata Institute of Fundamental Research
TMI	Three Mile Island

TOE	Ton of Oil Equivalent
UN	United Nations
UNSCEAR	United Nations Scientific Committee on the Effects of Atomic Radiation
UNSCOM	United Nations Special Commission
UOX	enriched Uranium OXide
USEC	United States Enrichment Corporation
VLLW	Very Low Level Waste
VVER	Russian abbreviation for "pressurised normal water reactor"
WEC	World Energy Council
WHO	World Health Organisation
W-Pu	Weapon Plutonium

Biographies

Nuclear Energy: Promise or Peril?

Executive Editor:

Bob van der Zwaan is a Physicist (CERN) and Economist (Cambridge) and has been a Nuclear Energy Researcher at the *Institut Français des Relations Internationales* (Paris).

Advisory Editors:

Kit Hill is Secretary of the British Pugwash Group and former Professor of Physics as Applied to Medicine at the Institute of Cancer Research, London.

André Mechelynck is Secretary of the Belgian Pugwash Group, a Consultant on Energy Matters and former Engineering Advisor.

Georges Ripka is President of the French Pugwash Group and Senior Physicist in the *Service de Physique Théorique* at the *Centre d'Etudes de Saclay*.

Authors:

Benjamin Dessus is Director of the Interdisciplinary Research Programme on Technologies for Eco-development "ECODEV" at the French CNRS.

Steve Fetter, Associate Professor, is Director of the Environmental Policy Program at the University of Maryland, US.

John Finney is Quain Professor of Physics, University College (London), and a former ISIS Chief Scientist at the Rutherford Appleton Laboratory.

David Fischer, a Consultant on Nuclear Non-proliferation and Safeguards, is a former Assistant Director General for External Relations at the IAEA (Vienna).

Richard Garwin is a Member of the US President's Science Advisory Committee, IBM Fellow Emeritus, and Adjunct Professor of Physics at Columbia University.

Pierre Goldschmidt is a Nuclear Energy and Plutonium Specialist, and General Manager, SYNATOM (Brussels).

Gert Harigel is Secretary of the Geneva International Peace Research Institute (GIPRI) and Senior Physicist Emeritus at CERN.

Jack Harris is Editor of the Interdisciplinary Science Reviews and former Senior Section Head at the Berkeley Nuclear Laboratories, UK.

Padmanabha Iyengar is a former Director of the Bhabha Atomic Research Centre (BARC) and Chairman of the Indian Atomic Energy Commission.

Georges le Guelte is a former Secretary of the IAEA Board of Governors and a former Deputy Director of International Relations at the French CEA.

Douglas Morrison is a former Senior Scientist and Honorary Staff Member at CERN, and Former Visiting Professor at the Universities of Vienna and Hawaii.

Bas Pease is Chairman of the British Pugwash Group and former Director of Fusion Research at the UK Atomic Energy Authority.

Jean-Paul Schapira, Nuclear Physicist, is Director of Research at the French *Centre National de la Recherche Scientifique* (CNRS).

Georges Vendryes is former Director of the *Applications Industrielles Nucléaires (Civiles)* Division of the French *Commissariat à l'Energie Atomique*.

Other Participants of the Review Workshop:

Eric Ash is Deputy President of the UK Royal Society and former Provost of Imperial College, London.

Eric Ferguson is a Consultant on Energy and Development Matters, and a former Senior Physicist at Philips Research.

Victor Gilinski is an Energy Consultant and a former Commissioner of the US Nuclear Regulatory Commission.

George Rathjens is Secretary General of the Pugwash Conferences on Science and World Affairs and Professor Emeritus at MIT.

Joseph Rotblat is the Joint 1995 Nobel Peace Prize Laureate, Past Pugwash President and Emeritus Physics Professor at St. Bartholomew's Hospital Medical School, London.

Etienne Roth is Honorary Professor at the *Conservatoire National des Arts et Metiers* (Paris) and Consultant at the *Centre d'Etudes Nucléaires de Saclay*.

Camille Sellier is a retired Rear-Admiral, and is Scientific Advisor to the French *Commissariat à l'Energie Atomique, Direction des Applications Militaires*.

Gérard Toulouse, Theoretical Physicist, is a Research Scientist at the *Ecole Normale Supérieure* in Paris and corresponding member of the *Académie des Sciences*.

Administrative Assistance:

Claudia Vaughn is working at the Rome Office of the Pugwash Conferences on Science and World Affairs.

Acknowledgements

The Pugwash Workshop on "The Prospects of Nuclear Energy", held at the French *Académie des Sciences* in Paris, on 4-5 December 1998, was organised by the Belgian, English and French national Pugwash groups.

Warm thanks are due to the following for funding the costs of the peer review Workshop, which played an essential part in the preparation of this book: the *Fondation de la Ferthé*, for a substantial donation, and the funds of the "Belgian Pugwash Group" and the "British Pugwash Trust".

Special thanks are expressed to Mr. Jean Dercourt, *Secrétaire Perpétuel* of the *Académie des Sciences*, for his words of welcome and for hosting the meeting.

Index[1]

E

F

G

H

I

V

W

Y

[1] Some words, occurring very often in this work (such as fission, (non-)proliferation, plutonium, reprocessing and uranium), have not been included in this index.